T And B Cell Cooperation In The Immune Response

Papers by
Fernando Morgado, G. Doria, Sara Kunin,
et al.

IN COOPERATION WITH THE
SMITHSONIAN SCIENCE INFORMATION EXCHANGE

Summaries of current research projects are included in the final section of this volume. Previously unpublished, these summaries were obtained from a search conducted by the Smithsonian Science Information Exchange, a national collection of information on ongoing and recently terminated research.

MSS Information Corporation
655 Madison Avenue, New York, N.Y. 10021

Library of Congress Cataloging in Publication Data
Main entry under title:

T and B cell cooperation in the immune response.

   1. T cells--Addresses, essays, lectures.
2.   B cells--Addresses, essays, lectures.
3.   Immunity--Addresses, essays, lectures.
I. Morgado, Fernando. [DNLM: 1. Immunity,
Cellular--Collected works. 2. B-Lymphocytes--
Immunology--Collected works. 3. T-Lymphocytes--
Immunology--Collected works. QW504 T11 1973]
QR185.8.L9T18   599'.02'9   73-12316
ISBN 0-8422-7116-3

TABLE OF CONTENTS

CREDITS AND ACKNOWLEDGEMENTS

Allison, A.C.; and A.J.S. Davies, "Requirement of Thymus-Dependent Lymphocytes for Potentiation by Adjuvants of Antibody Formation," *Nature*, 1971, 233:330-331.

Andersson, J.; G. Möller; and O. Sjöberg, "B Lymphocytes Can Be Stimulated by Concanavalin A in the Presence of Humoral Factors Released by T Cells," *European Journal of Immunology*, 1972, 2:99-101.

Basten, A.; J. Sprent; and J.F.A.P. Miller, "Receptor for Antibody-Antigen Complexes Used To Separate T Cells from B Cells," *Nature New Biology*, 1972, 235:178-180.

Cheers, Christina; J.C.S. Breitner; Margery Little; and J.F.A.P. Miller, "Cooperation between Carrier-reactive and Hapten-sensitive Cells *in Vitro*," *Nature New Biology*, 1971, 232:248-249.

Dennert, G.; and E. Lennox, "Cell Interactions in Humoral and Cell-Mediated Immunity," *Nature New Biology*, 1972, 238:114-116.

Doria, G.; G. Agarossi; and S. di Pietro, "Effect of Blocking Cell Receptors on an Immune Response Resulting from *in Vitro* Cooperation between Thymocytes and Thymus-Independent Cells," *The Journal of Immunology*, 1971, 107:1314-1318.

Doria, G.; G. Agarossi; and S. di Pietro, "Enhancing Activity of Thymocyte Culture Cell-Free Medium on the *in Vitro* Immune Response of Spleen Cells from Neonatally Thymectomized Mice to Sheep RBC," *The Journal of Immunology*, 1972, 108:268-270.

Greaves, Melvyn F.; and Sara Bauminger, "Activation of T and B Lymphocytes by Insoluble Phytomitogens," *Nature New Biology*, 1972, 235:67-70.

Janossy, G.; and M.F. Greaves, "Lymphocyte Activation. I. Response of T and B Lymphocytes to Phytomitogens," *Clinical and Experimental Immunology*, 1971, 9:483-498.

Kunin, Sara; Gene M. Shearer; Shraga Segal; Amiela Globerson; and Michael Feldman, "A Bicellular Mechanism in the Immune Response to Chemically Defined Antigens. III. Interaction of Thymus and Bone Marrow-Derived Cells," *Immunology*, 1971, 2:229-238.

Lonai, Peter; and Michael Feldman, "Cooperation of Lymphoid Cells in an *in Vitro* Graft Reaction System. II. The 'Bone Marrow-Derived' Cell," *Transplantation*, 1971, 11:446-456.

Mitchison, N.A., "The Carrier Effect in the Secondary Response to Hapten-Protein Conjugates," *European Journal of Immunology*, 1971, 1:10-17.

Morgado, Fernando; and Hugh Folch, "Limited Synergism Between Rat Thymus and Mouse Bone Marrow," *Nature*, 1971, 232:637.

Niederhuber, J.; Erna Möller; and O. Mäkelä, "Cytotoxic Effect of Anti-Θ and Anti-Mouse Specific B Lymphocyte Antigen (MBLA) Antisera on Helper Cells and Antibody-Forming Cell Precursors in the Immune Response of Mice to the 4-hydroxy-3,5-dinitrophenacetyl (NNP) Hapten," *European Journal of Immunology*, 1972, 2:371-374.

Orsini, Frank R.; and Gustavo Cudkowicz, "Thymic Antigen-Reactive Cells Do Not Specify Serological Properties of Antibody," *Cellular Immunology*, 1971, 2:300-308.

Playfair, J.H.L., "Response of Mouse T and B Lymphocytes to Sheep Erythrocytes," *Nature New Biology*, 1972, 235:115-118.

Raff, Martin C., "T and B Lymphocytes in Mice Studied by Using Antisera against Surface Antigenic Markers," *The American Journal of Pathology*, 1971, 65:467-476.

Roelants, G.E., "Quantification of Antigen Specific T and B Lymphocytes in Mouse Spleens," *Nature New Biology*, 1972, 236:252-254.

Schimpl, A.; and E. Wecker, "Replacement of T-Cell Function by a T-Cell Product," *Nature New Biology*, 1972, 237:15-17.

Sulitzeanu, D.; R. Kleinman; D. Benezra; and I. Gery, "Cellular Interactions and the Secondary Response *in Vitro*," *Nature New Biology*, 1971, 229:254-255.

Vann, Douglas C.; and John R. Kettman, "*In Vitro* Cooperation of Cells of Bone Marrow and Thymus Origins in the Generation of Antibody-Forming Cells," *The Journal of Immunology*, 1972, 108:73-80.

Waksman, B.H.; M.C. Raff; and June East, "T and B Lymphocytes in New Zealand Black Mice: An Analysis of the Theta, TL and MBLA Markers," *Clinical and Experimental Immunology*, 1972, 11:1-11.

Wilson, J.D.; and Marc Feldmann, "Dynamic Aspects of Antigen Binding to B Cells in Immune Induction," *Nature New Biology*, 1972, 237:3-5.

Wioland, M.; D. Savolovic; and C. Burg, "Electrophoretic Mobilities of T and B Cells," *Nature New Biology*, 1972, 237:274-276.

PREFACE

The immune response represents the sum total of the body's reaction to a foreign agent. Sorting out the many components of this response is a problem now facing investigators in several areas of research. One aspect of immunology which has raised great interest is the role played by various populations of immunocompetent cells.

It is now clear that at least two types of cells interact to provide an antibody response. There is also strong evidence which suggests that cellular immune responses and immunological unresponsiveness also require collaboration between various populations of lymphocytes. The lymphocytes involved in these reactions have been defined as thymus dependent (T cells) and thymus independent (B cells). The T cells require the thymus for maturation, while B cells -- though derived from the bone marrow -- require other undefined sites for maturation into competent lymphocytes.

Evidence for T and B cell interactions results from three sources. Early experiments involving the X-irradiation of mice and their protection from damage by the injection of allogeneic bone marrow provided the first source. These animals were shown to lack an immune response unless thymic lymphocytes were also injected. Thus, it was indicated that T and B cells are required for immune responsiveness. A multitude of studies have since pointed out the complexity of these reactions. Secondly, immunolgoical deficiency syndromes in man also suggested a compartmentalization of lymphoid development and function. These studies, coupled with those using birds, demonstrated that T and B cells develop under the influence of different microenvironments and that both cell types are necessary for complete immunocompetence. Lastly, phylogenetic studies have further emphasized nature's delegation of immunologic functions to many cell types.

In preparing this compilation of recent studies on the nature of T and B cell collaboration, I have attempted to include literature dealing with both the early concepts and more recently evolved ones. Research involving general aspects of these phenomena have been selected as well as that dealing with more precise examples of cooperation in cellular and humoral responses. Most exciting are those papers which indicate that a soluble factor may be secreted by the T cells, thereby directly influencing the B cells to respond to an antigen. Reports on new techniques which enable a more direct means of identifying and separating the T and B cell populations are also presented. These have all been chosen to provide conceptual direction for future attempts to manipulate the immune mechanism for the benefit of mankind.

Ronald T. Acton, Ph.D.
June, 1973

Example of Thymus and Bone Marrow Cell Cooperation

Limited Synergism between Rat Thymus and Mouse Bone Marrow

Fernando Morgado
Hugo Folch

In a number of experimental systems[1] it has been shown that thymus and bone marrow-derived cells can cooperate in the production of humoral antibodies. In some instances this synergism has been demonstrated when the interacting cells are antigenically similar but in other experiments allogeneic cells have been found to cooperate. We have now demonstrated cooperation between rat and mouse cells. A two-step experiment was performed, following the design of Miller[2]. Rat or mouse thymus cells were introduced with antigen into irradiated recipients. The donor cells were collected and injected into secondary host mice. The details of these experiments and their results are as follows.

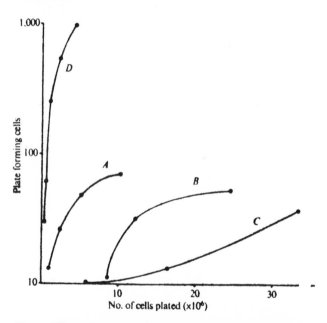

Fig. 1 Effect of mouse and rat thymus on the immunocompetence of mouse bone marrow cells (see text). Group A lethally irradiated mice repopulated with mouse thymus and mouse bone marrow cells; group B, lethally irradiated mice repopulated with rat thymus and mouse bone marrow cells; group C, lethally irradiated mice repopulated with mouse bone marrow cells; group D, unirradiated control mice given only SRBC.

Twenty adult RK mice, 10–12 weeks old, were given 900 r. of total body irradiation from a ^{60}Co source. Ten were injected with 2×10^7 mouse thymus cells plus 0.1 ml. of SRBC at 20% (group 1), and ten were injected with 2×10^7 rat thymus cells plus 0.1 ml. of SRBC at 20% (group 2).

Five days later another twelve RK mice, 10–12 weeks old, were given total body irradiation, and then three were injected with $1 \cdot 5 \times 10^6$ cells from group 1 pooled spleens plus 2×10^7 mice bone marrow cells and 0.1 ml. of SRBC at 20% (group A), three were injected with $1 \cdot 5 \times 10^6$ cells from group 2 pooled spleens plus 2×10^7 mice bone marrow cells and 0.1 ml. of SRBC at 20% (group B), and three were injected with 2×10^7 mice bone marrow cells and 0.1 ml. of SRBC at 20% (group C). As controls three irradiated and three non-irradiated mice (group D) were immunized with 0.1 ml. of SRBC at 20%.

Six days later we counted plaque forming cells against SRBC—of pooled spleens of the different groups—by the Jerne technique[3]. All cellular suspensions were prepared in cold phosphate buffer saline, P 7.4, and injected intravenously within 3 h of the irradiation.

Fig. 1 shows that rat and mouse cells can cooperate, although not as efficiently as mouse-mouse combinations. The mechanism of cooperation is not understood but if it involves recognition of organ specific surface antigens it seems that these may be recognized across genetic barriers.

[1] Claman, H. N., and Chaperon, E. A., *Transplant. Rev.*, 1, 92 (1969).
[2] Miller, J. F. A. P., and Mitchell, G. F., *Transplant. Rev.*, 1, 3 (1969).
[3] Jerne, N. K., and Nordin, A. A., *Science*, 140, 405 (1963).

EFFECT OF BLOCKING CELL RECEPTORS ON AN IMMUNE RESPONSE RESULTING FROM *IN VITRO* COOPERATION BETWEEN THYMOCYTES AND THYMUS-INDEPENDENT CELLS[1]

G. DORIA, G. AGAROSSI AND S. DI PIETRO

The immune capacity of spleen cells from neonatally thymectomized mice to produce antibodies to sheep RBC *in vitro* was restored by adding thymocytes to culture. This cell cooperation was prevented when the spleen cells but not thymus cells were treated with rabbit antibodies to mouse serum proteins prior to culture. Thymocytes appeared to lack demonstrable receptor molecules such as those detected on thymus-independent cells and present in normal serum.

Several observations have provided experimental evidence for the existence of receptor sites on immunologically competent cells. The ability of lymphoid cell populations to bind antigen (1, 2) and the presence of Ig-like molecules on lymphocyte surface membranes (3–5) suggest that immunologically competent cells recognize antigen by membrane-attached Ig. This view is supported by the finding that the antigen binding of normal (6) or primed (7) lymphoid cells can be blocked by preincubating the cells with antibodies directed against serum Ig. Whether the Ig involved in antigen recognition contains light or heavy chains of a known class or both has been investigated to some extent. Antigenic stimulation of primed cells for a humoral response could be blocked *in vitro* by addition of antibodies against IgG Fab (8). The reaction of human lymphocytes to PPD or allogenic antigens (9) and the ability of mouse spleen or peritoneal exudate cells to induce graft-*vs*-host reactions or to transfer delayed hypersensitivity (10) could be suppressed by cell treatment with antibodies against Ig light chains. Similarly, pretreatment of normal mouse spleen cells with anti-Ig light chain antibodies (11) or with anti-kappa chain and complement (12) inhibited primary antibody production *in vivo* or *in vitro*, respectively. With regard to heavy chains, normal mouse cells pretreated with antibodies to μ, α, γ_1 or γ_2 chains were assayed for competence in antigen binding or antibody production (11) and

[1] Supported by CNEN-Euratom Association Contract. Publication No. 717 of the Euratom Biology Division.

in the graft-*vs*-host reaction (10). Suppression of activity was observed only in this last case when antibodies to μ chains had been used. However, mouse antibody-forming or memory cells treated with antisera prior to filtration through antigen-coated columns had a surface receptor with the same kappa and γ_1 or γ_{2a} chains as those of the secreted antibody (13).

Peripheral lymphoid tissues are made up of mixed populations of thymus-dependent and thymus-independent lymphocytes. Although both cell types derive from the bone marrow, the former originates through cellular differentiation under thymic influence, whereas the latter develops directly from the bone marrow. *In vivo* experiments demonstrated that thymus-dependent and thymus-independent cells are immunologically competent lymphocytes, both involved in several immune responses. Thus, in the mouse immunized with sheep red blood cells (RBC), specific antibodies are produced by thymus-independent cells upon interaction with thymus-dependent cells and the antigen (14). The nature of this interaction is not understood. It has been suggested that the thymus-dependent cell binds antigen by receptor sites specific for some antigenic determinants and focuses other determinants of the same antigen molecule onto specific receptors of the thymus-independent cell. Hence, the former cell traps antigen so as to form a local concentration at some critical site on the latter cell, which is then triggered to antibody synthesis (15).

Cooperation between thymocytes from normal mice and splenocytes from neonatally thymectomized mice or from thymectomized chimeras (mice thymectomized in adult life, lethally irradiated, and grafted with isogenic bone marrow cells) was shown to occur *in vitro*, as a primary immune response could be induced *in vitro* only if both the thymocytes and thymus-independent cells were cultured with sheep RBC (16). This *in vitro* system was considered suitable for investigating whether receptor sites are involved in the interaction between spleen cells from neonatally thymectomized and thymus cells from normal mice. The present work deals with receptor molecules shared by cells and normal serum. Thus, pretreatment of mouse cells with a rabbit antiserum against mouse serum prior to culture may block receptors and prevent cell function.

MATERIALS AND METHODS

Animals. (C57BL/10 ♀ × DBA/2 ♂)F_1 hybrid mice of either sex were used in all experiments.

In vitro cell cooperation. The Mishell and Dutton technique (17) was used whereby unprimed mouse spleen cells were stimulated *in vitro* with sheep RBC to give rise to hemolytic plaque-forming cells (PFC). Cell preparations and culture conditions were the same as described previously (16). Cells were obtained from 45-day-old normal or neonatally thymectomized (18) mice. In each experiment 1-ml fractions of separate or mixed cell suspensions were distributed in four groups of culture dishes as follows: a) 1.5×10^7 nucleated spleen cells from normal mice; b) 2.5×10^7 nucleated spleen cells from thymectomized mice; c) 2.5×10^7 nucleated thymus cells from normal mice; d) 1.5×10^7

nucleated spleen cells from thymectomized and 1×10^7 nucleated thymus cells from normal mice. Antigen (1×10^7 sheep RBC/dish) was added to each group. Control cultures received no antigen. From day 4 on, cells were harvested daily from two dishes of the same group, pooled, washed and assayed for the number of direct PFC by the Jerne technique (19).

Inhibition of immune response. Prior to culture, spleen cells from thymectomized mice and thymus cells from normal mice were separately treated as follows. One volume of packed cells was suspended with an equal volume of balanced salt solution (BSS) (17), rabbit normal serum (NS) or rabbit antiserum (AS) against mouse serum and allowed to stand at 4°C for 4 hr except for being gently shaken every 15 min. Thereafter, the cells of each suspension were washed twice with 100 volumes of BSS and, after nucleated cell counting, appropriate volumes of the two suspensions were mixed according to the treatment combinations specified under *Results*. One-milliliter portions of the mixed cell suspensions were distributed in culture dishes, each of which received 1.5×10^7 nucleated spleen cells, 1×10^7 nucleated thymus cells and 1×10^7 sheep RBC. In each mixed culture the two cell populations had been subjected to the same or different treatment. Also spleen cells from normal mice were treated with BSS, NS or AS prior to culture (1.5×10^7 nucleated cells and 1×10^7 sheep RBC/dish) as described above.

The antiserum was prepared as follows. Four rabbits were given three intramuscular injections, every 3 weeks, of 0.4 ml of normal mouse whole serum emulsified in 0.4 ml of complete

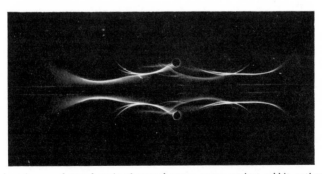

Figure 1. Agar immunoelectrophoresis of normal mouse serum against rabbit antiserum against mouse whole serum.

13

Freund's adjuvant. From the third week after the last injection, the rabbits were given weekly six intravenous injections of 0.4 ml of mouse serum. The animals were bled from the heart 1 week after the last injection. Sera were collected and tested by immunoelectrophoresis against normal mouse serum. The pattern of the AS selected for cell treatment is shown in Figure 1. Before use, both NS and AS were heated at 56°C for 30 min. and sterilized by filtration.

RESULTS

Figure 2 shows the results of a typical cell cooperation experiment. The addition of sheep RBC to mixed cell cultures of thymus cells from normal mice and spleen cells from thymectomized mice elicited an immune response comparable with that of normal spleen cells. When the two cell populations were stimulated in separate cultures no response of thymus cells was ever

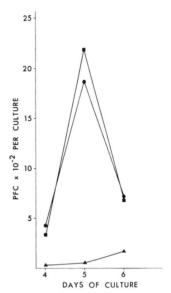

Figure 3. Inhibition of immune response. Spleen cells from normal mice were pretreated with BSS ●——●, NS ■——■, or AS ▲——▲. Sheep RBC were added to all cultures.

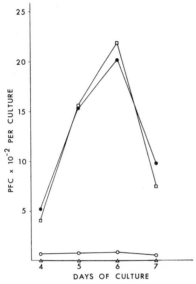

Figure 2. Immune response resulting from *in vitro* cell cooperation. Spleen cells from normal mice ●——●, spleen cells from neonatally thymectomized mice ○——○, thymus cells from normal mice △——△, spleen cells from neonatally thymectomized mice and thymus cells from normal mice □——□. Sheep RBC were added to all cultures.

observed, whereas the response of spleen cells from thymectomized mice was in nearly all cases within the range of the PFC background in unstimulated control cultures.

The effect of cell pretreatment with BSS, NS or AS on the immune response of normal spleen cells is shown in Figure 3. The drastic inhibitory action of AS is apparent. The slight enhancement of the response of cells pretreated with NS was consistently observed in several experiments of this kind.

The effects of cell pretreatments on the immune response in mixed cultures of cooperating spleen and thymus cells are illustrated in Table I. The results demonstrate that the response could be inhibited only when the spleen cells were pretreated with AS. Incubation of thymus cells with AS before antigenic stimulation did not interfere with the cell cooperation events leading to antibody formation.

DISCUSSION

The results of the present experiments with $(C57BL/10 \times DBA/2)F_1$ mice confirm and ex-

TABLE I

Effect of cell pretreatment on immune response (PFC/culture) in mixed cultures

Exp. No.	Cell Pretreatment		Days of Culture			
	Tx-Spleen[a]	Thymus[b]	4	5	6	7
1	BSS[c]	BSS	380	330	750	453
	NS[d]	BSS	460	750	910	830
	AS[e]	BSS	109	120	40	16
2	NS	BSS	475	1245	1405	490
	AS	BSS	58	45	373	200
3	BSS	BSS	320	3798	1470	467
	BSS	NS	325	3473	710	275
	BSS	AS	337	2855	1100	176
4	BSS	BSS	159	180	1230	380
	BSS	NS	215	480	1630	940
	BSS	AS	86	110	1480	240
5	NS	NS	480	920	1165	125
	NS	AS	213	690	1135	300
	AS	NS	8	80	245	48
	AS	AS	7	10	15	5
6	NS	NS	68	268	1213	
	NS	AS	325	985	3235	
	AS	NS	33	15	30	
	AS	AS	68	60	443	
7	NS	NS	825	1455	5895	
	NS	AS	630	3383	3915	
	AS	NS	83	373	1680	
	AS	AS	33	405	775	

[a] Tx-Spleen = spleen cells from thymectomized mice.
[b] Thymus = thymus cells from normal donors.
[c] BSS = balanced salt solution.
[d] NS = rabbit normal serum.
[e] AS = rabbit antiserum against mouse serum.

tend our previous demonstration that thymus cells can help isogenic bone marrow-derived cells to produce antibodies in vitro. The source of bone marrow-derived cells was the spleen either from neonatally thymectomized DBA/2 mice or from C3HeB/Fe thymectomized chimeras (16). In similar in vitro systems, spleen cells from normal CBA mice (20) or spleen cells from (C57BL/6 × DBA/2)F$_1$ mice irradiated and injected with thymus cells and antigen (21) were shown to cooperate with isogenic spleen cells from thy-

mectomized chimeras. In these studies, however, thymus cells only occasionally were able (20) or failed (21) to restore in vitro the immune response of spleen cells from thymectomized chimeras. At present it is difficult to resolve this discrepancy with our results, as neither the strain differences nor the different sources of bone marrow-derived cells seem to be convincing explanations.

As shown in Figure 3 and Table I, treatment of spleen cells from normal or thymectomized mice with rabbit antiserum for 4 hr followed by extensive cell washing prior to culture markedly inhibited the immune response in vitro. No appreciable differences were ever observed among numbers of nucleated cells recovered from cultures of cells pretreated with BSS, NS or AS. Since PFC hardly exceeded 0.1 % of the recovered nucleated cells, numeric variations of PFC were not expected to be detected by nucleated cell counts. Thus, inhibition of the response may be attributed to specific action on immunologically competent cells rather than to nonspecific cell toxicity of the rabbit antiserum.[2] Furthermore, the latter possibility seems to be ruled out by the inefficiency of the same antiserum on thymus cells.

Lesley and Dutton (12) found inhibition of the in vitro immune response of spleen cells from normal (C57BL/6 × DBA/2)F$_1$ mice by pretreating cells with rabbit antiserum and complement or by adding the antiserum to culture. However, cell pretreatment at 37°C for 1 hr with the same antiserum alone had no inhibitory effect. This disagreement with our results showing inhibition by cell pretreatment with rabbit antiserum alone may be explained by the different incubation times and temperatures. In fact, looking for optimal experimental conditions at 4°C, we found 4 hr of incubation much more efficient than 0.5, 1 or 2 hr in depressing the in vitro immune response of normal spleen cells.

Impairment of the immune response of normal spleen cells following pretreatment with rabbit antiserum against mouse whole serum indicates that splenic lymphoid cell populations possess receptor sites antigenically identical or similar to

[2] Viability cell counts were considered unnecessary, for nucleated cell counts performed on four consecutive days of culture should have revealed cell toxicity of AS as compared with that of NS and BSS.

15

molecules of the normal serum. As impairment was irreversible, it may have resulted from permanent blockade of cell receptors or from cell death. In the cell cooperation experiments, inhibition was observed when spleen but not thymus cells had been incubated with rabbit antibodies. Thus, cell receptors appear to be localized on thymus-independent cells, as it has not been possible to detect on thymus cells receptor sites sharing antigenic determinants with molecues of the normal serum. The use of polyvalent rather than anti-Ig rabbit antiserum allows one to extend this conclusion to serum molecules antigenically different from Ig. Thus, either receptors are not present on thymus cells or, if they are present, they are not secreted into the serum in sufficient concentration to be immunogenic when injected in rabbits as described. In the latter case, thymocyte receptors ought to have a different structure from those on thymus-independent cells, unless a different steric arrangement on the thymus cell membrane prevented the same receptor molecules from combining with rabbit antibodies.

The present findings are in agreement with both the lack of demonstrable Ig on thymocytes (5, 22–24) and the thymocyte ability to bind antigen in a specific way (25–27). In fact, thymocytes were not prevented by rabbit antibodies to mouse Ig (and other serum proteins) from interacting with antigen and thymus-independent cells. This *in vitro* system should allow further analysis of the mechanisms by which the thymocytes bind antigen and trigger thymus-independent cells to antibody synthesis.

REFERENCES

1. Naor, D. and Sulitzeanu, D., Nature, *214:* 687, 1967.
2. Wigzell, H. and Andersson, B., J. Exp. Med., *129:* 23, 1969.
3. Sell, S. and Gell, P. G. H., J. Exp. Med., *122:* 423, 1965.
4 Coombs, R. R. A., Feinstein, A. and Wilson, A. D., Lancet, *2:* 1157, 1969.
5. Raff, M. C., Sternberg, M. and Taylor, R. B., Nature, *225:* 553, 1970.
6. Byrt, P. and Ada, G. L., Immunology, *17:* 503, 1969.
7. Biozzi, G., Binaghi, R. A., Stiffel, C. and Mouton, D., Immunology, *16:* 349, 1969.
8. Mitchison, N. A., Sympos. Quant. Biol., *32:* 431, 1967.
9. Greaves, M. F., Torrigiani, G. and Roitt, I. M., Nature, *222:* 885, 1969.
10. Mason, S. and Warner, N. L., J. Immun., *104:* 762, 1970.
11. Warner, N. L., Byrt, P. and Ada, G. L., Nature, *226:* 942, 1970.
12. Lesley, J. and Dutton, R. W., Science, *169:* 487, 1970.
13. Walters, C. S. and Wigzell, H., J. Exp. Med., *132:* 1233, 1970.
14. Miller, J. F. A. P. and Mitchell, G. F., Transplant. Rev., *1:* 3, 1969.
15. Mitchison, N. A., in *Immunological Tolerance*, Edited by M. Landy and W. Braun, Academic Press, New York, 1969.
16. Doria, G., Martinozzi, M., Agarossi, G. and Di Pietro, S., Experientia, *26:* 410, 1970.
17. Mishell, R. I. and Dutton, R. W., J. Exp. Med., *126:* 423, 1967.
18. Miller, J. F. A. P., Brit. J. Cancer, *14:* 93, 1960.
19. Jerne, N. K. and Nordin, A. A., Science, *140:* 405, 1963.
20. Munro, A. and Hunter, P., Nature, *255:* 277, 1970.
21. Hartmann, K. U., J. Exp. Med., *132:* 1267, 1970.
22. Pernis, B., Forni, L. and Amante, L., J. Exp. Med., *132:* 1001, 1970.
23. Rabellino, E., Colon, S., Grey, H. M. and Unanue, E. R., J. Exp. Med., *133:* 156, 1971.
24. Paraskevas, F., Lee, S.-T. and Israels, L. G., J. Immun., *106:* 160, 1971.
25. Dwyer, J. M. and Mackay, I. R., Lancet, *1:* 1199, 1970.
26. Modabber, F. and Coons, A. H., Fed. Proc., *29:* 697, 1970.
27. De Luca, D., Decker, J. and Sercarz, E. E., Fed. Proc., *29:* 697, 1970.

A Bicellular Mechanism in the Immune Response to Chemically Defined Antigens

III. Interaction of Thymus and Bone Marrow-Derived Cells [1]

Sara Kunin, Gene M. Shearer , Shraga Segal,
Amiela Globerson, and Michael Feldman

The role of thymus and bone marrow-derived cells in the *in vitro* response to the dinitrophenyl (DNP) determinant was studied using the millipore filter well technique for spleen organ cultures. Antibodies to DNP were assayed by the technique of inactivation of DNP-coupled T-4 bacteriophage. It was found that spleens of mice total-body irradiated at 750 R, treated with bone marrow and thymus cells after exposure and immunized against rabbit serum albumin (RSA) were able to produce antibodies to DNP when challenged *in vitro* with DNP–RSA. Such a response was not produced by spleen explants from x-irradiated mice treated with either thymus or bone marrow cells. Neither were antibodies to DNP produced by spleens of animals repopulated with thymus and bone marrow cells, but not immunized with the carrier. This carrier effect was manifested when the irradiated mice were treated with RSA and thymus cells 6–8 days before administration of the bone marrow cells. Yet, such an effect was not observed when the RSA and bone marrow cells were given 6–8 days before injection of the thymus cells. Thus, the thymus-derived cells appear to play the role of cells sensitive to the carrier (RSA), whereas the bone marrow seems to be involved in the production of antibodies.

INTRODUCTION

The notion that antibody production is based on a bicellular interaction between thymus and bone marrow-derived cells stemmed from experiments with SRBC as an antigen (1–3). In this system evidence was brought that each of the two interacting cells, which differ ontogenically, differ also in function: the thymus-derived cell functions as an antigen-sensitive cell, i.e., as a cell which recognizes antigenic determinants but does not produce antibodies, whereas the bone marrow-derived cell both recognizes and is capable of actual production of antibodies (4, 5). The production of antibodies to hapten determinants, following immunization with hapten

[1] This work was supported by a grant from the Max and Ida Hillson Foundation, New York, and from DGRST, France.

carrier conjugates, was also claimed to be based on bicellular cooperation. This was first inferred from the observation that an anamnestic response to a hapten is achieved only if the carrier used for the secondary stimulation is identical to the one applied in the primary immunization with the hapten carrier conjugate (6). The interpretation of these results on the basis of cooperation between cells recognizing carrier determinants and cells recognizing hapten determinants gained further support from our experiments on the *in vitro* production of antibodies to the dinitrophenyl determinant (DNP) (7, 8). These indicated that the primary response to a hapten is determined by a mechanism of bicellular cooperation since (a) either free DNP or free carrier molecules can prevent production of anti-DNP antibodies; (b) primary immunization against the carrier alone increases the response to the DNP upon subsequent stimulation with hapten carrier conjugates.

The present study aimed at: (a) testing directly whether antibody production to DNP is determined by cooperation between thymus and bone marrow-derived cells, and (b) testing the role of thymus and bone marrow-derived cells in terms of recognition of the antigenic determinants of the carrier and the hapten.

We employed the *in vitro* system in which we previously demonstrated the production of a primary response to chemically defined antigens in cultures (9), and in which we studied the nature of the cell receptors for antigenic determinants functioning in the primary and in the secondary response (8). Spleen explants tested *in vitro* in the present investigation were obtained from x-irradiated animals which were reconstituted with thymus and bone marrow cells. Such reconstituted spleens were employed by us for the analysis of cell cooperation in the *in vitro* production of antibodies to sheep red blood cells (SRBC) (10).

MATERIALS AND METHODS

Mice. Male (Balb/c x C57Bl/6)F_1 mice, 8–10 weeks old, were used throughout these experiments.

Irradiation. Mice were exposed to 750 R total-body irradiation using a Picker-Vanguard x-ray machine (250 kV peak, 15 mA) with 0.5 mm Cu and 1.0 mm Al added filtration. They were irradiated at a target distance of 50 cm (exposure rate of 60 R/min).

Culture technique. The millipore filter-well technique for the induction of antibody response *in vitro* was employed as previously described for sheep red blood cells (SRBC) (11) and was adapted for chemically defined antigens (9).

Immunogens. The dinitrophenyl (DNP) determinant was used as the hapten in this study. The carrier employed was rabbit serum albumin (RSA), fraction V, obtained from Nutritional Biochemical Co. DNP was attached to the carrier at a molecular ratio of 5:1, respectively (designated as DNP–RSA).

Immunization. (a) *In vivo:* Mice received a single intraperitoneal injection of 0.2 mg of RSA or DNP–RSA in complete Freund's adjuvant. (b) *In vitro:* Spleen explants were incubated with 0.01 ml of medium containing 50 μg of DNP–RSA/ml. Forty-eight hrs later the culture medium was replaced by antigen-free medium.

Antibody assay. Culture media collected on the 7th to 9th day of culture were assayed for anti-DNP antibodies by the modified T-4 bacteriophage (2,4-dinitrophenyl bacteriophage T-4, DNP-T4) technique (12).

18

Cell transfer. Thymus, bone marrow, and spleen cells for transfer were prepared from 6–8-weeks-old mice as described elsewhere (13). Cells were injected intravenously immediately after x-ray exposure, at inoculum size of 50×10^6 spleen, 35×10^6 bone marrow and 10×10^7 thymus cells pere recipient. At certain time intervals after transfer the recipients' spleens were removed and were explanted as previously described (10).

EXPERIMENTAL

We first aimed at testing whether the production of antibodies to DNP is determined by an interaction between thymus and bone marrow-derived cells. The experimental design was based on the notion that cells originating in the thymus and in the bone marrow may achieve immune reactivity only after colonizing lymphoid organs, such as the spleens (14). We, therefore, planned to test the capacity to produce in culture antibodies to DNP by spleen explants of animals which were subjected to lymphoid cell depletion by total-body x-irradiation, and then were experimentally repopulated with thymus and bone marrow cells. To examine the applicability of this approach based on cell transfer, we first tested in culture spleens of animals exposed to total-body x-irradiation and then inoculated with spleen cells from normal donors. The following experiment was made: Irradiated mice were inoculated with 50×10^6 spleen cells from untreated syngeneic donors immediately after x-ray exposure, and were injected with RSA 24 hr later. Six days after irradiation, the mice were sacrificed and fragments of their spleens were cultured *in vitro*. One group of cultures was incubated with DNP–RSA, and a second group of nontreated cultures served as control. Cultures of spleens from animals which were not immunized against RSA were set up in parallel.

It was found that only cultures of spleens originating in mice immunized against RSA and stimulated *in vitro* with DNP–RSA produced a significant level of antibodies to the hapten (Table 1, Fig. 1). Thus, the "carrier effect" can be obtained in this *in vitro* system using spleens from irradiated mice repopulated with lymphoid cells.

An experiment was then performed to establish whether induction of a primary response to a hapten involves participation of bone marrow and of thymus-derived

TABLE 1

In Vitro Response to DNP–RSA by Spleens of Irradiated Mice Injected with Spleen Cells and Immunized Against the Carrier (RSA)

Antigen injected to spleen donors	No. of spleens tested	Antigen added *in vitro*	Mean % inactivation [a] of DNP-T4
RSA		DNP–RSA	81.5 (5) [b]
RSA	10	—	37.4 (5)
—	10	DNP–RSA	23.6 (5)

[a] Medium samples were collected on day 8, pooled and diluted 1:3.

[b] Number of cultures per spleen. Mean percentage values were calculated from the results of the 10 different pools, containing five cultures each.

19

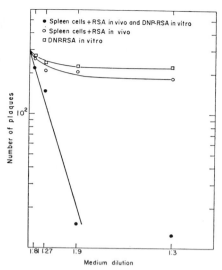

• Spleen cells + RSA in vivo and DNP-RSA in vitro
○ Spleen cells + RSA in vivo
□ DNP-RSA in vitro

Number of plaques

10^2

Medium dilution

FIG. 1. In vitro response to DNP–RSA by spleens of irradiated mice injected with spleen cells and immunized against the carrier (RSA). Medium samples were collected on day 8 from one representative experiment, pooled and assayed at different dilutions for inactivation of DNP-T4 bacteriophage.

cells as required for more complex antigens. Irradiated mice were divided into three groups. One group was inoculated with 10×10^7 thymus cells together with 35×10^6 bone marrow cells, the second group was injected with bone marrow, and the third with thymus cells only. Twenty-four hours later all three groups were injected with DNP–RSA. Recipient mice were sacrificed 10 days after immunization and fragments of their spleens were explanted *in vitro*. Cultures of each group were divided into two subgroups: one was inoculated with DNP–RSA and a second group without antigen served as control. Medium samples were assayed on the 8th day of culture for the presence of antibodies to DNP. It was found that antibodies were produced only in the first group of cultures, i.e., in those originating from recipients injected with thymus and bone marrow cells and stimulated *in vitro* with the antigen (Table 2, Fig. 2). No response was detected in the groups of either thymus or bone marrow-treated donors. Thus, the production of antibodies to DNP *in vitro* seems to require participation of thymus and bone marrow-derived cells.

What is the specific role played by each of these cell types? Previous observations on the carrier effect (8, 15) have shown that the response to DNP involves cells with receptors to the hapten as well as cells capable of recognizing carrier determinants. Accordingly, the carrier-sensitive cells could be the cells of thymus origin which were shown to act as antigen sensitive cells (ASC) in other experimental systems employing more complex antigens, e.g., SRBC (1–3) and proteins (16). We, therefore, planned to determine which of the two cell populations, bone marrow or thymus-derived cells, act as carrier-sensitive cells.

TABLE 2

In Vitro Response to DNP–RSA by Spleens of Irradiated Mice Injected with Thymus and/or Bone Marrow Cells and Immunized Against DNP–RSA

Cells transferred	No. of spleens tested	Mean % inactivation of DNP-T4 [a]	
		Antigen added *in vitro*	Control
Thymus and bone marrow	4	97.4 (10) [b]	19.4 (10) [b]
Thymus	4	0 (10)	0 (10)
Bone marrow	4	11.0 (10)	19.8 (10)

[a] Medium samples were collected on day 8, pooled and diluted 1:3.

[b] Number of cultures per spleen. Mean percentage values were calculated from the results of the four different pools, containing 10 cultures each.

Irradiated mice were divided into three groups. The first group was inoculated with thymus and bone marrow cells, the second with bone marrow only, and the third with thymus cells only. Twenty-four hours later they were injected with RSA. The recipient mice were sacrificed 10 days later, explants of their spleens were prepared and challenged *in vitro* with DNP–RSA. As demonstrated in Table 3 and Fig. 3, a significant level of response was obtained only in the first group of cultures originating in mice which had received both bone marrow and thymus

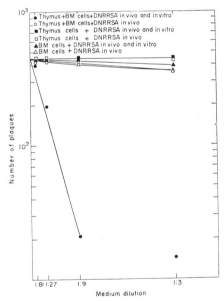

Fig. 2. *In vitro* response to DNP–RSA by spleens of irradiated mice injected with thymus and/or bone marrow cells and immunized against DNP–RSA. Medium samples were collected on day 8 from one representative experiment, pooled and assayed at different dilutions for inactivation of DNP-T4 bacteriophage.

TABLE 3

In Vitro Response to DNP–RSA by Spleens of Irradiated Mice Injected with Thymus and/or Bone Marrow Cells and Immunized Against the Carrier (RSA)

Cells transferred [a]	No. of spleens tested	Mean % inactivation of DNP-T4 [b]	
		Cultures stimulated *in vitro* with DNP–RSA	Control
Thymus and bone marrow	8	96.6 (10) [c]	0 (10) [c]
Thymus	8	26.7 (10)	36.2 (10)
Bone marrow	8	29.5 (10)	24.0 (10)

[a] Spleens were explanted 10 days following irradiation.

[b] Medium samples were collected on days 8 and 9, pooled and diluted 1:3.

[c] Number of cultures per spleen. Mean percentage values were calculated from the results of the eight different pools containing 10 cultures each.

cells. Spleens treated with either bone marrow or thymus cells alone did not produce anti-DNP antibodies under these experimental conditions.

In order to determine which of the two cell populations contains the carrier-sensitive cells, we decided to expose selectively one of these cell populations to the carrier before interaction with the second cell population occurred. Thus, irradiated mice were first inoculated either with thymus cells or with bone marrow cells.

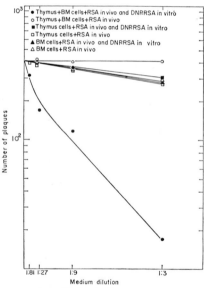

Fig. 3. *In vitro* response to DNP–RSA by spleens of irradiated mice injected with thymus and/or bone marrow cells and immunized against the carrier (RSA). Medium samples were collected on day 8, from one representative experiment, pooled and assayed at different dilutions for inactivation of DNP-T4 bacteriophage.

Twenty-four hours later the recipients were injected with RSA. The second cell population was injected 6 days later, 4 days before sacrificing the animals. Spleen cultures were divided into two groups: one was stimulated with the immunogen, DNP–RSA, and the second served as an unstimulated control. As shown in Table 4 and Fig. 4, only spleen cultures originating in donors which were first inoculated with thymocytes and RSA produced antibodies in response to stimulation with DNP–RSA *in vitro*. This experiment was repeated with a different schedule of time intervals by using 8 days between first and second transfer of cells and 2 days between second transfer and explantation. The results were similar to those obtained in the previous experiment (Table 4, Expt. 2 and Fig. 4). Hence, thymus-derived cells appear to play the role of carrier-sensitive cells in induction of a response to DNP.

DISCUSSION

Results of the present study demonstrate that the immune response to DNP coupled to RSA, as well as the carrier effect of preimmunization with RSA, can be obtained by repopulation of heavily irradiated mice either with suspensions of syngeneic spleen or mixtures of thymus and bone marrow cells, but not by bone marrow or thymus cells alone. These findings are in agreement with the observations of others that cell-to-cell cooperation is required for the immune response to haptens (17, 18), and are consistent with the conclusion that the carrier effect and subsequent enhanced antibody production to the hapten is mediated by a mechanism involving at least two functionally distinct cell types (8, 19). Of the im-

TABLE 4

In Vitro Response to DNP–RSA by Spleens of Irradiated Mice Injected with Thymus and Bone Marrow Cells at Different Time Intervals and Immunized Against the Carrier (RSA) 24 hr after First Cell Transfer

				Mean % inactivation of DNP-T4 [b]	
Expt. No.	Cells transferred	No. of spleens tested	Explantation time (days) [a]	Cultures stimulated *in vitro* with DNP–RSA	Control
1	Thymus 6 days before bone marrow cells	6	4	96.7 (5) [c]	25.9 (5) [c]
	Bone marrow 6 days before thymus cells	6	4	22.9 (5)	39.5 (5)
2	Thymus 8 days before bone marrow cells	6	2	88.1 (5)	33.2 (5)
	Bone marrow 8 days before thymus cells	6	2	23.3 (5)	4.1 (5)

[a] Time interval between second injection and explantation of spleen.

[b] Medium samples were collected on day 8, pooled and diluted 1:3.

[c] Number of cultures per spleen. Mean percentage values were calculated from the results of the six different pools, containing five cultures each.

munogens thus far tested, most appear to be "thymus dependent" in the sense that antibody production in lethally irradiated mice appears to require both thymus and bone marrow-derived cells from donor animals. One reported exception is polymerized *Salmonella* flagellin, which elicits an immune response in marrow-reconstituted mice without the apparent help of thymocytes (20). It is not known at the present time whether the cell cooperation demonstrated for the carrier effect is a feature of a "thymus-dependent" carrier, or whether in fact a "thymus-independent" antigen can give a carrier effect at all.

The experiments involving the inoculation of thymocytes and marrow cells separated by a time interval, were similar in design to experiments of others using SRBC (21, 22). In our case, however, the experiments were performed to test in which cell population the exposure to RSA was necessary in order to obtain the carrier effect. It was found that the injection of thymocytes and RSA followed by bone marrow cells 6–8 days later, was as effective in generating the carrier effect as the simultaneous injection of a thymus–bone marrow cell mixture. On the other hand, the initial injection of bone marrow and RSA followed by an injection of thymocytes 6–8 days later, did not generate a carrier effect above that observed for control cultures. These results indicate that the carrier effect induced by RSA was primarily on the thymocyte population and not on the relevant immunocompetent precursors of bone marrow origin. It should be noted that the

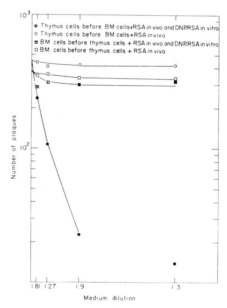

Fig. 4. *In vitro* response to DNP–RSA by spleens of irradiated mice injected with thymus and bone marrow cells at different time intervals and immunized against the carrier (RSA) 24 hr after first cell transfer. Medium samples were collected on day 8 from one representative experiment, pooled and assayed at different dilutions for inactivation of DNP-T4 bacteriophage.

preimmunization of marrow cells with SRBC 4–6 days before the introduction of other cooperating cell types was reported to have no detectable priming effect (21, 22), whereas a strong priming effect was observed when the time interval was 8 days (22). The absence of a carrier effect in bone marrow is unlikely to be due to an insufficient period of immunization, since both 6 and 8 day intervals were tested.

The precise effect of pretreatment with the carrier on the thymus-derived lymphocytes is not clear. Since production of antibodies is a function of bone marrow-derived cells, yet its realization required the coexistence of both bone marrow and specific thymus-derived cells, preimmunization with RSA might have stimulated the replication of thymus-originated cells with specific receptors for RSA. Upon subsequent application of both bone marrow cells and DNP–RSA, the probability of functional cooperation via the immunogen between RSA-specific thymus-derived cells and bone marrow-derived DNP-specific cells is enhanced. Such an enhanced probability of cooperation could be envisaged also if the carrier caused an increase in the number of RSA receptors per RSA-sensitive cell, rather than an increase in cell number. Experiments using inhibitors of cell replication have indeed suggested that the "carrier effect" might be based on stimulating the production of more receptors per cell, since blocking cell proliferation 24 hr following the application of RSA, did not prevent the carrier effect manifested upon a subsequent application of DNP–RSA (7, 8). Thus, thymus-derived cells stimulated with RSA could have responded by synthesizing more cell receptors for RSA, a process which might require just one cycle of cell division.

ACKNOWLEDGMENTS

The excellent technical assistance of Miss Miriam Kotler is gratefully acknowledged.

REFERENCES

1. Davies, A. J. S., *Transplant. Rev.* 1, 43, 1969.
2. Miller, J. F. A. P., and Mitchell, G. F., *Transplant. Rev.* 1, 3, 1969.
3. Claman, H. N., and Chaperon, E. A., *Transplant. Rev.* 1, 92, 1969.
4. Davies, A. J. S., Leuchars, E., Wallis, V., Marchant, R., and Elliot, E. V., *Transplantation* 5, 222, 1967.
5. Miller, J. F. A. P., and Mitchell, G. F., *J. Exp. Med.* 131, 675, 1970.
6. Ovary, Z., and Benacerraf, B., *Proc. Soc. Exp. Biol. Med.* 114, 72, 1963.
7. Segal, S., Kunin, S., Globerson, A., Shearer, G. M., and Feldman, M., *In* "Procedures of the Third Congress on Lymphatic Tissue and Germinal Centers," Plenum Press, New York (in press).
8. Segal, S., Globerson, A., and Feldman, M., *Cell. Immunol.* (in press).
9. Segal, S., Globerson, A., Feldman, M., Haimovich, J., and Sela, M., *J. Exp. Med.* 131, 93, 1970.
10. Globerson, A., and Feldman, M., *In* "Mononuclear Phagocytes," Blackwell Scientific Publications (R. van Furth, ed.), p. 613, 1970.
11. Globerson, A., and Auerbach, R., *J. Exp. Med.* 124, 1001, 1966.
12. Haimovich, J., and Sela, M., *J. Immunol.* 97, 338, 1966.
13. Shearer, G. M., and Cudkowicz, G., *J. Exp. Med.* 129, 935, 1969.
14. Feldman, M., and Globerson, A., *In* "Immunogenicity" (F. Borek, ed.), North Holland Publishing Co., Amsterdam (in press).

15. Segal, S., Globerson, A., and Feldman, M., *Cell. Immunol.* (in press).
16. Taylor, R. B., *Transplant. Rev.* **1**, 114, 1969.
17. Mitchison, N. A., *Cold Spring Harbor Symp. Quant. Biol.* **32**, 431, 1967.
18. Rajewsky, K., Schirrmacher, V., Nase, S., and Jerne, N. K., *J. Exp. Med.* **129**, 1131, 1969.
19. Katz, D. H., Paul, W. E., Goidl, E. A., and Benacerraf, B., *J. Exp. Med.* **132**, 261, 1970.
20. Armstrong, W. D., Diener, E., and Shellam, G. R., *J. Exp. Med.* **129**, 393, 1969.
21. Shearer, G. M., and Cudkowicz, G., *J Exp Med.* **130**, 1243, 1969.
22. Kennedy, J. C., Treadwell, P. E., and Lennox, E. S., *J. Exp. Med.* **132**, 353, 1970.

Cellular Interactions and the Secondary Response *in Vitro*

D. Sulitzeanu   D. Benezra

R. Kleinman   I. Gery

ALTHOUGH the role of cellular cooperation in the induction of the immune response has become firmly established only recently[1], morphological evidence suggesting that such cooperation takes place is quite old. Reports[2] of the aggregation of lymphoid cells around macrophages[3] have been confirmed: "islets", "rosettes" or "clusters" in cultures of cells (derived from humans[4], guinea-pigs[5], rabbits[6] or mice[7]) stimulated with antigen or PHA[8] were reported. We have investigated cluster formation to ascertain its relationship, if any, to the antigen-induced stimulation of sensitized cells[9]. We used peripheral blood leucocytes from rabbits immunized to bovine serum albumin (BSA) or to human red blood cells (HRBC). The BSA was given in complete Freund's adjuvant (three intramuscular injections of 7.5 mg BSA each, into the hind legs at weekly intervals). HRBC (1 ml.) was given once into the ear vein, as a 20% suspension in saline. Cell cultures and ^{3}H-thymidine incorporation were measured as before[10]. To prepare cell smears, cells were washed three times and suspended in one drop of normal rabbit serum and 1 μl. of the suspension was spread on a microscope slide. This ensured a reasonably constant number of cells per slide and made possible comparisons between different experiments. Smears were fixed with methanol and stained with Giemsa.

Clusters were counted 2–5 days after cultures were set up. A group of cells was considered a "cluster" if it consisted of a large central leucocyte surrounded by at least three lymphoid cells. The extent of cluster formation was expressed as the cluster index—the number of clusters (from a mean count of two to four slides) per slide of stimulated culture divided by the number of clusters per slide of non-stimulated culture. Cluster index values were somewhat larger if calculated on the basis of counts of larger clusters (five peripheral cells or more) rather than on the basis of total cluster counts. The results of ^{3}H-thymidine incorporation were expressed as the incorporation index—c.p.m. in the stimulated cultures divided by c.p.m. in the non-stimulated cultures. Autoradiograms were prepared with Ilford K5 nuclear emulsion to study ^{3}H-thymidine incorporation and ^{125}I-BSA uptake. ^{3}H-Thymidine (2.5 µCi; 1.9 Ci/mmol) was added to the culture tubes 2 h before collection, or cultures were incubated for various times with labelled BSA (2 µCi; specific activity 50 µCi/µg) mixed with 1 µg of unlabelled BSA.

Table 1 presents typical results obtained in fourteen experiments, which involved eighty-eight counts of clusters. Antigen-stimulated cultures contained more clusters than non-stimulated cultures and the difference was usually larger with the larger clusters (five peripheral cells or more). Two types of central cells were seen; one was obviously a macrophage (Fig. 1a, b, c and e) with pale cytoplasm and an irregular nucleus, and the other had a large rounded nucleus and basophilic cytoplasm,

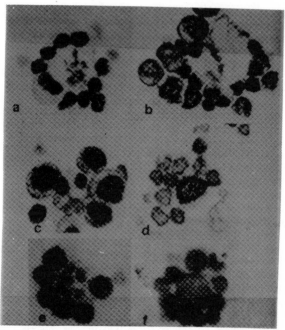

Fig. 1 Macrophage clusters: a, cluster consisting mostly of peripheral small lymphocytes; b, larger cluster (taken from an experiment with spleen cells) in which many of the peripheral cells undergo blast transformation; c, macrophage cluster in which most of the peripheral cells have taken up ³H-thymidine; d, blast type clusters. The central cell in d has incorporated ³H-thymidine; e and f, autoradiography of cells from rabbits immunized to BSA, stimulated in vitro with radioiodinated BSA, with antigen in central cells; e, macrophage cluster; f, blast type cluster. (× 480.)

suggesting a blast form (Fig. 1d and f). The peripheral cells varied in size from true small lymphocytes (Fig. 1a) to large cells apparently undergoing blastoid transformation (Figs. 1b and c). More of the small lymphocytes were seen towards the end of the incubation. Cultures of normal cells generally contained few clusters, which did not increase after stimulation.

The specificity of cluster formation was tested by stimulating with BSA, HRBC or SRBC cultures of cells from rabbits immunized with either BSA or HRBC. ^{3}H-Thymidine incorporation and cluster formation increased only after stimulation with the antigen to which the rabbit had been immunized.

To compare the kinetics of cluster formation with those of ^{3}H-thymidine incorporation, cultures from three donors immunized to BSA were stimulated with 10 or 1,000 µg of BSA. After 2, 3 and 5 days, cluster indices and incorporation indices were determined and cell smears were subjected to autoradiography. After 2 days all cluster indices were larger than incorporation indices (in all of six determinations), whereas after 5 days incorporation indices were larger than cluster indices in five out of six determinations. Evidently therefore cluster formation precedes thymidine incorporation. Autoradiographic examination revealed that many of the central blast-like cells in the clusters had strongly labelled nuclei (Fig. 1d). By contrast, the nuclei of the typical macrophages did not take up thymidine. In some of the cultures, pulse labelled after 5 days, occasional macrophage clusters had one or more labelled blast like cells at the periphery (Fig. 1c). Autoradiography of cells from cultures stimulated with ^{125}I-BSA showed that many of the central cells in the clusters, both of the macrophage and of the blast-like type, contained antigen, while cells at the periphery were free of radioactivity (Fig. 1e and f).

These results suggest that contact of lymphoid cells with macrophages and blast cells in clusters is the first stage in the transformation leading to the appearance of antibody forming cells. The evidence for this is quite impressive: (1) *in vitro* stimulation is enhanced by increasing the cell density[4], which presumably facilitates cell–cell interaction, but is depressed by shaking the cultures[11]; (2) stimulation is inhibited if cell–cell contact is prevented mechanically[6]; (3) macrophages are needed for the transformation to take place[4]; (4) cluster formation precedes ^{3}H-thymidine incorporation; (5) cells with labelled ^{3}H-thymidine can be found among the peripheral cells in clusters (ref. 8 and this communication); (6) clusters contain most of the antibody forming cells[7]; (7) the secondary response to antigen is accompanied by increased cluster formation.

The blast-like cells found in the stimulated cultures, which can interact with antigen and are stimulated to multiply, are fulfilling, in effect, two of the requirements for memory cells. If they are memory cells, our findings could be interpreted as follows: memory cells reacting with antigen are "activated", begin to multiply and, by an unknown mechanism, exert a tactic effect on other lymphoid cells, leading to the formation of clusters. Any cells in the clusters capable of responding to

Table 1 Cluster Formation and ³H-Thymidine Incorporation After Stimulation of BSA Sensitized Cultures with Antigen

| Stimulating dose (μg) | Days of incubation | Cluster formation | | | | ³H-Thymidine incorporation | |
| | | No. of clusters | | Total | Cluster index | c.p.m. | Incorporation index |
		Small*	Large†				
—	3	62	33	95		126	
		97	62	159	1.8	518	4.1
1,000		112	112	224	3.4	1,172	9.3
—	5	68	31	99		272	
		43	37	80	1.1	682	2.5
1,000		87	45	132	1.4	3,732	13.7

* Small clusters consisting of a central cell surrounded by three to four peripheral cells.
† Large clusters consisting of a central cell surrounded by five or more peripheral cells.

antigen would presumably be stimulated to produce antibodies (as suggested by the association of antibody forming cells with clusters in the primary response[7] and by several reports[12] of the apparent antibody formation by secondarily stimulated cell cultures). Antigen laden macrophages probably have a similar tactic effect, causing the formation of macrophage clusters and leading to the activation of "antigen reactive cells" or "X" cells. Thus, whereas "blast clusters" might be a morphological representation of the secondary response, "macrophage clusters" could represent the induction of a primary response *in vitro*. This interpretation fits the current idea that the three cell types[13] (macrophage, antigen reactive cell and antibody forming cell) require cluster formation[7] for the induction of a primary response.

This work was supported in part by the Concern Foundation, Los Angeles, and by the Samuel Lautenberg fellowship. We thank Dr G. Rasooli for his help.

[1] Miller, J. F. A. P., and Mitchell, G. F., *Transplantation Rev.*, **1**, 3 (1969).
[2] Thiery, J. P., *J. Microsc.*, **1**, 275 (1962).
[3] McFarland, W., and Heilman, D. H., *Nature*, **205**, 887 (1965).
[4] Hersh, E. M., and Harris, J. E., *J. Immunol.*, **100**, 1184 (1968).
[5] Bartfeld, H., and Kelly, R., *J. Immunol.*, **100**, 1000 (1968).
[6] Harris, G., *Immunology*, **9**, 529 (1965).
[7] Mosier, D. E., *J. Exp. Med.*, **129**, 351 (1969).
[8] Richter, M., and Naspitz, C. K., *Blood*, **32**, 134 (1968).
[9] Sulitzeanu, D., Rasooli, G., Benezra, D., and Gery, I.. *Israel J. Med. Sci.*, **5**, 443 (1969).
[10] Gery, I., Eyal, O., and Benezra, D., *Immunology* (in the press).
[11] Richardson, M., Moorhead, J. W., and Reedy, D. L., *Immunology*, **17**, 603 (1969).
[12] Girard, J. P., *Intern. Arch. Allergy*, **32**, 294 (1967).
[13] Mosier, D. E., and Coppleson, L. W., *Proc. US Nat. Acad. Sci.*, **61**, 542 (1968).

The carrier effect in the secondary response to hapten-protein conjugates.

N. A. Mitchison

Abbreviations: AFCP: Antibody-forming-cell-precursors   CGG: Chicken γ-globulin   BSA: Bovine serum albumin   HSA: Human serum albumin   BGG: Bovine γ-globulin   OA: Ovalbumin   KLH: Keyhole limpet hemocyanin   MGG: Mouse γ-globulin   NIP: 4-Hydroxy-5-iodo-3-nitro-phenacetyl   DNP: 2,4-Dinitro-phenyl   NP-CAP: N^ϵ-(4-hydroxy-3-nitro-phenacetyl-) amino-caproic-acid   NIP-CAP: N^ϵ-(4-hydroxy-5-iodo-3-nitro-phenacetyl-) amino-caproic acid   HOP-DNP-lys: N^α-(4-hydroxy-phenacetyl-) N^ϵ-(2,4-dinitrophenyl-lysine   DIP-DNP-lys: N^α-(3,5-diiodo-4-hydroxy-phenacetyl-)-N^ϵ-(2,4-dinitrophenyl-)lysine. l. u.: log units (see section 2.5)

I. Measurement of the effect with transferred cells and objections to the local environment hypothesis

A carrier effect is obtained typically when a hapten-protein conjugate is injected into an animal which has previously been primed with the same hapten conjugated to another carrier protein. Under these circumstances the anti-hapten secondary response is usually less than that which would have been obtained had the animal been injected with a conjugate prepared with the same carrier as that originally used for priming. Attempts have been made to account for the phenomenon in terms of the local environment hypothesis, which assumes that the receptor on immunologically competent cells recognises the hapten jointly with the area on the complete antigen which surrounds it. Alternatively the phenomenon can be accounted for by the hypothesis of cooperation, which assumes that the antigen is recognised by two receptors, one directed to the hapten and the other to a determinant on the carrier protein.

32

Methods are described which enable carrier effects to be studied quantitatively in mice. They involve a cell transfer system in which cell suspensions prepared from large numbers of donors are distributed among irradiated syngeneic recipients. In these recipients the transferred cells can be made to perform a secondary response by appropriate antigenic stimulation. The response is monitored by binding tests in which the capacity of serum to bind highly radioactive haptens or proteins is measured. The haptens employed in this system are NIP (4-hydroxy-5-iodo-3-nitro-phenacetyl-) and DNP (2,4-dinitrophenyl-) and the proteins comprise chicken γ-globulin, bovine serum albumin, human serum albumin, ovalbumin, bovine γ-globulin, keyhole limpet hemocyanin and mouse γ-globulin.

A carrier effect was regularly obtained when the proteins were tested against one another as carriers, for priming and for the secondary response. The effect could best be measured by comparing the relative potencies in the secondary response of the homologous conjugate (*i. e.* one with the carrier which had originally been used for priming) with heterologous conjugates (*i. e.* ones with new carriers). In this way the intrinsic potency of the individual protein could also be measured and allowance made for it in calculating the magnitude of the carrier effect. An average carrier effect of one thousand-fold relative potency was obtained. Priming by NIP-ovalbumin or NIP-chicken γ-globulin with secondary stimulation by NIP-bovine serum albumin (or the corresponding DNP conjugates) could be identified as the combination best suited to further study.

Support for the cooperation hypothesis, particularly for that version of the hypothesis which postulates that recognition of carrier determinants allows an antigen-concentrating mechanism to operate, could be found in the parallel slopes of the dose-response curves obtained with homologous and heterologous conjugates. On the other hand the local environment hypothesis failed to pass either of the tests to which it was subjected. One, the weaker, was to compare haptens with and without spacer groups inserted between themselves and the carrier protein, in the expectation that spacers might reduce the local environment contribution: no difference could in fact be detected. The other, the stronger, was to attempt to inhibit the response with an excess of carrier protein even though the anti-hapten antibody had no detectable affinity for the carrier: such inhibition could regularly be obtained.

1. Introduction

This series of papers is concerned with an interaction between two populations of lymphocytes which plays an important part in the control of the immune response. The investigation starts from an examination of the secondary response of mice to hapten-protein conjugates. It goes on to show that the carrier effect can be explained by an interaction between two cell populations, one of which provides AFCP and the other helper cells. Evidence will then be presented that the helper cell in this response is a thymus-derived lymphocyte and that the interaction between helpers and AFCP therefore maps onto the thymus-murrow synergy which has been the subject of much recent work [1, 2, 3, 4]. It will be shown that findings obtained with the usual combi-

nation of AFCP-anti-hapten and helper-anti-carrier apply to other species and other combinations of determinants. The following interpretation of the helper-AFCP interaction will be offered: helper cells pick up antigen via specific receptors which combine with one set of determinants; this increases the effective local concentration of other antigenic deter- minants in the vicinity of the receptors on AFCP, which are consequently triggered.

Carrier effects [5,6] occur in the induction of the secondary response, [7,8] in the induction of immunological tolerance [9] and in eliciting delayed hypersensitivity [10]. Typically a hapten conjugated to one carrier protein fails to elicit an anti-hapten response in an individual previously immunized by the same hapten conjugated to another carrier. The effect, however, is general and not confined to hapten-protein com- binations [11, 12, 13].

The first coherent explanation offered of the carrier effect was that the receptor responsible for induction of the anti- hapten response recognised an area of the antigen surrounding the hapten — "the local environment" — and failed to recog- nise the hapten in a different local environment (for illustra- tion, see [14]). Closely allied to this hypothesis is the suppo- sition that the heterogeneity of the anti-hapten response can be attributed to variation in the local environment of the hap- ten on the immunogen [15, 16]. Opposed to it is the hypo- thesis of cooperation [11, 17, 18, 19] which postulates the recognition of antigen by two receptors, one directed to the hapten and the other to a determinant on the carrier protein.

This paper examines quantitatively the secondary response of mice to a range of hapten-protein conjugates, with the prin- cipal aim of defining conditions under which the carrier ef- fect can best be studied. It also includes a number of expe- riments designed to test predictions of the local environment hypothesis. Brief accounts of this work and of that reported in the subsequent papers of this series have already been published [6, 17, 20, 21, 22, 23]

2. Materials and Methods

2.1. Animals and Immunization

Male mice of the CBA strain were used. They normally entered use at 10—12 weeks at a weight of 25—30 g.

Mice were immunized except where specified by intraperitoneal injection of 400—800 μg protein or conjugate adsorbed onto alum, mixed with 2 x 10^9 killed *Bordetella pertussis* organisms in a final volume of 0.2 ml [24, 25]. To adsorb onto alum, one part 2 % protein was mixed with one part 9 % KAl $(SO_4)_2 \cdot 12 H_2O$, the pH raised to 6.5 with 2N NaOH, and the resulting precipitate washed twice with pH 7.2 phosphate-buffered saline. Alum precipitates were stored at 2 °C with the addition of 1/50 000 merthiolate. Mice were used for secondary response tests 1—4 months after immunization, except where specified.

2.2. Cell transfers

Up to 60 donor mice were killed by cervical dislocation and the skin over the left flank snipped with scissors and peeled back with the fingers. Then, under clean but non-sterile conditions, the spleens were collected into ice-cold balanced salt solution and all subsequent manipulations were carried on in this medium at 0—4 °C. In earlier experiments Gey's solution was used, but in later experiments this was replaced by bicarbonate-free balanced salt solution (1 g dextrose, 0.06 g KH_2PO_4, 0.358 g Na_2HPO_4, 0.01 g phenol red plus 0.186 g $CaCl_2 \cdot 2 H_2O$, 0.4 g KCl, 8.0 g NaCl, 0.2 g $MgCl_2 \cdot 6 H_2O$, 0.2 g $MgSO_4 \cdot 7 H_2O$ per l). The spleens were gently pressed through a stainless steel screen, 10 wires/cm, and then passed by syringe through a 16 g needle once or twice to free cells. The suspension was allowed to settle for a few min and the scraps of tissue which settled out were discarded. The suspension was washed once on the centrifuge (300 x g, 10 min) and then injected intraperitoneally. One spleen consistently yielded 120 x 10^6 cells of which more than 50 % were viable as judged by trypan blue exclusion. Except where specified each host received 40 x 10^6 spleen cells (total nucleated cells).

One day before cell transfer the prospective hosts were irradiated with 600 r (in earlier experiments from a 140 kV X-ray source, in later experiments from ^{60}Co γ radiation). One day after cell transfer the hosts were injected intraperitoneally with 0.2 ml antigen, in the dose specified, and 8—10 days later they were bled from the tail and sera collected.

2.3. Antigens

The following proteins were used in this and subsequent papers in this series: (1) CGG prepared by fractionation (45 % saturated ammonium sulfate) of fresh slaughterhouse chicken serum ($E_{280\,nm}^{1\%,1\,cm}$ = 14, MW = 160 000); (2) BSA from Armour, crystallized for serology, otherwise Cohn fraction V ($E_{280\,nm}^{1\%,1\,cm}$ = =6.8; MW = 69 000); (3) HSA crystallized from Behringwerke ('reinst') ($E_{280\,nm}^{1\%,1\,cm}$ = 5.8; MW = 69 000); (4) OA from Worthington ($E_{280\,nm}^{1\%,1\,cm}$ = 7.4; MW = 45 000); (5) BGG from Armour ($E_{280\,nm}^{1\%,1\,cm}$ = 14; MW = 160 000), for serology purified by passage through DEAE-cellulose (0.02 M phosphate pH 7.0); (6) KLH from Mann (MW = 8 x 10^6); (7) MGG prepared by fractionation (45 % saturated ammonium sulfate, DEAE-cellulose, 0.02 M phosphate pH 7.0) of mouse serum ($E_{280\,nm}^{1\%,1\,cm}$ = 13; MW = 160 000).

NIP-conjugates were prepared [26] by reacting protein (2–5 %) dissolved in ice-cold 0.5 M NaHCO$_3$ (pH 8) with a solution of 4-hydroxy-3-iodo-5-nitro-phenacetyl azide (NIP azide) in dimethylformamide so that the final concentration of dimethylformamide in water did not exceed 5 %. The mixture was stirred overnight at 2 °C and then dialysed exhaustively (4 changes in 48 h) against 0.2 M NaHCO$_3$ and finally saline. Under these conditions 40–80 % of the theoretical yield of coupled NIP was obtained. Coupling ratios were estimated from the molar extinction of NIP-CAP of 5.0 x 10^3 at 430 nm [26].

DNP-conjugates were prepared similarly using a pH 9.5 carbonate buffer and reacting with 2,4-dinitrobenzene sulphonic acid. Coupling ratios were estimated on the basis of a molar extinction of DNP of 17 x 10^3 at 360 nm.

Conjugates with spacer groups were prepared as follows. N^{11}-(4-hydroxy-3-iodo-5-nitrophenacetyl) undecanoic acid azide [26] was reacted with protein under the conditions specified for NIP azide to yield NIP-N^{11}-conjugates. NIP-ala$_n$ refers to conjugates prepared by reacting NIP azide with polyalanylated carriers. The poly-D L-alanyl BSA (478 additional alanines per molecule of BSA) and poly-D L-alanyl OA (545 additional alanines per molecule of OA) used for this purpose were gifts from Dr. S. Bauminger of the Weizman Institute. NIP-ala$_4$ was prepared by reacting NIP azide with tetra-L-alanine (Yeda) according to the method previously used to prepare NIP-gly-gly [26]; coupling with protein was then brought about by means of water soluble carbodiimide. Thus 60 mg BSA was reacted with 10 mg NIP-ala$_4$ in the

36

presence of 75 mg N-(cyclohexyl-iminomethylenaminoethyl)-N-methyl-morpholinium-p-toluene-sulfonate in 4 ml dilute NH_4OH to yield $(NIP\text{-}ala_4)_{5.8}$ BSA.

All conjugates had fairly low coupling ratios (5–15 haptens per molecule protein) except where specified.

Proteins and conjugates were stored frozen at a concentration > 5 mg/ml and when diluted to $< 100\ \mu g$/ml protective protein was added (rabbit serum or, when this might cross-react, mouse serum).

2.4. Serology

Haptens with high specific radioactivity for use in binding tests were prepared by radio-iodination of NP-CAP to yield N^{125}IP-CAP [26], and by radio-iodination of HOP-DNP-lys to yield D^{125}IP-DNP-lys [17, 27]. Thus, for example, to 1 mCi carrier-free ^{125}I (Amersham, IMS. 3) was added in succession 25 μl 0.3 M phosphate buffer pH 7.2, 10 μl 2×10^{-3} M NP-CAP + 2×10^{-3} KI, 25 μl chloramine-T (2 mg/ml) and 100 μl $Na_2S_2O_5$(2 mg/ml). The mixture was then transferred onto a small column (1 cm in a Pasteur pipette) of ion-exchange resin (Dowex AG 1 x 8, chloride form, 200–400 mesh, Biorad) and rinsed with 1 ml water.

The N^{125}IP-CAP was then eluted with 3 x 0.5 ml 50 % acetic acid and the acetic acid immediately neutralised with 2N NaOH. All the ^{125}iodide was retained on the column and the yield was 60 % with respect to ^{125}I.

D^{125}IP-DNP-lys was prepared in exactly the same manner, except that the product could not easily be eluted from ion-exchange resin and so ^{125}iodide was removed on a Sephadex G-25 (Pharmacia) column. Elution was carried out with pH 8.4 borate buffer: the ^{125}iodide emerged in the first peak (1 x included volume) and the D^{125}IP-DNP-lys in the second (1.8 x included volume).

Proteins were also radio-iodinated by the chloramine-T method, using 25 μl of a solution containing 1 mg protein/ml, and the same quantity of the other reagents. The ^{125}iodide was separated on Sephadex G-25 columns.

For use in binding tests N^{125}IP-CAP was diluted in borate buffer (6.18 g H_3BO_3, 9.54 g $Na_2B_4O_7 \cdot 10\ H_2O$, 4.38 g NaCl per l) to a concentration sufficient to yield 10^{-8} M after addition to the serum under test and was adjusted with NIP-CAP to approximately 10 000 counts per 40 sec (the standard counting time) per tube. $D^{125}IP$-CAP was similarly diluted and adjusted with unlabeled DIP-CAP. Proteins were diluted in borate buffer containing 1 % protective normal rabbit serum to yield a final concentration of 1 μg/ml ($\cong 10^{-8}$M) and were adjusted with non-radioiodinated protein.

Sera were stored 1/6 in borate buffer at -2 °C until large numbers were ready for titration together. Then for salt precipitation (Farr) tests the sera were tested either at this 1/6 dilution only for weak antisera, or further dilutions of 1/36 and 1/180 in 10 % normal rabbit serum were prepared by means of a metering pump and Lang-Levy pipette. To 0.25 ml of the diluted serum in a 9 x 50 mm test-tube was added 0.25 ml of the radioactive hapten or antigen. The tubes we-were mixed and left to stand for a few hours, then 0.5 ml saturated ammonium sulfate was pumped in and the tubes were shaken and centrifuged (3 000 x g, 30 min, 4 °C). The supernatants were decanted, using in the process a wire screen to retain 42 tubes together in a centrifuge adapter. The precipitates were then counted in a well-type γ spectrometer without further washing, together with one set of control tubes containing either 0.25 ml of the radioactive hapten originally added or a TCA precipitate in the case of protein antigens (high controls) and with another set containing the ammonium sulfate precipitate of the radioactive hapten or protein mixed with 0.25 ml 10 % normal rabbit serum (low control). This procedure was used with the radioactive haptens and with BSA, HSA and OA (with OA the ammonium sulfate was adjusted to reach a final concentration of 40 %).

For CGG, BGG and KLH, the binding test employed rabbit anti-mouse IgG rather than ammonium sulfate to separate bound from free antigen. The anti-IgG serum was passed through columns of Sepharose-bound CGG or BGG [28] where necessary to remove cross-reactive antibody. Dilutions of antiserum were set up either at 1/6 only, or also at three further serial six-fold dilutions in 1/24 normal mouse serum in borate buffer by means of an Arnold hand microapplicator (Burkard). To 0.05 ml of the diluted serum in a polystyrene precipitation tube was added 0.01 ml of radioactive antigen. The tubes were mixed and sufficient anti-mouse IgG added to precipitate all available mouse antibody (usually 25 μl). After overnight incubation at 2° C, 0.2 ml phosphate buffered saline was added and the tubes mixed and centrifuged (3 000 x g, 30 min). The supernatants were sucked off and the precipitates counted, together with high and low controls analogous to those used in the salt precipitation test.

2.5 Calculations and statistical analysis

Readings from the salt precipitation tests were calculated as described in detail previously [25]. The formula previously reported for calculating the binding capacity contains an error and should correctly be written:
When binding > 35 %, therefore antigen binding capacity

$$= 2 \times 1/10 \text{ antilog} \left(\frac{\% \text{ binding}}{35} \times \log 3.5 \right) \bigg/ \text{dilution.}$$ Readings

from the anti-mouse IgG tests were calculated in exactly the same way when single tube tests were performed. When serial dilutions were used the binding capacity was estimated from the interpolated point at which 50 % binding occured. Three alternative calculations were thus used, all of which were carried out by computer using Fortran programs.

Because the variance of the binding capacities did not depend on titre when the capacities were measured on a log scale [25, 29], all measurements are expressed in the present series of papers either as $\log_{10}$ binding capacity in Figures, or as log units in Tables. Log units (1. u.) are based on the $\log_{10}$ scale; 1 l. u. (anti-hapten antibody) = 10^{-7} M hapten binding capacity of serum at a free hapten concentration of 10^{-8} M; 1 l. u. (anti-protein antibody) = $10 \mu g/ml$ antigen binding capacity of serum at a free antigen concentration of $1 \mu g/ml$. In the secondary response 1 l. u. hapten binding capacity $\cong 5 \mu g$ antibody protein [30].

All groups started with 6 mice and each data point therefore refers to the $\log_{10}$ mean of 6 (or less if animals died) values. Standard deviations are not given (they fell normally in the range 0.2–0.5 l. u.), but T-tests are calculated where appropriate.

3. Results

3.1. Measurement of the carrier effect

Some examples of the responses of primed mouse cells to a wide range of doses of antigen are given in Fig. 1 and others can be found in previous publications, or elsewhere

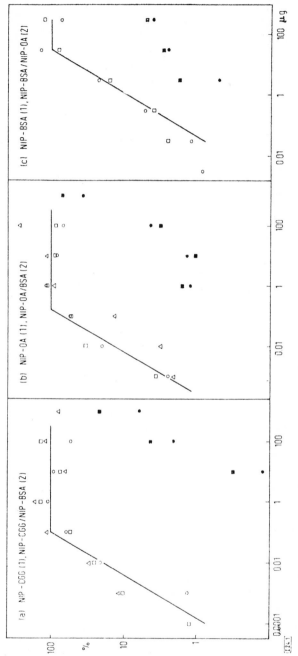

Figure 1. Anti-NIP responses to homologous (open symbols) and heterologous conjugates (closed symbols); in each figure the data for a single experiment are marked as a circle, square, or triangle. All experiments are adjusted to the same plateau and the responses are expressed as a percentage of the plateau. The responses are measured as a function of dose of antigen (μg/mouse). In (a) cells were taken from donors primed (1) with NIP-CGG and were stimulated (2) either with NIP-CGG (homologous conjugate) or NIP-BSA (heterologous conjugate). In (b) the homologous conjugate is NIP-OA and the heterologous NIP-BSA. In (c) the homologous conjugate is NIP-BSA and the heterologous NIP-OA.

in the present series of papers. These all follow the same pattern. Normally no antibody can be detected without antigenic stimulation, so that there is a clear base-line. Only in exceptional cases, attributed to unmasking of sequestered antigen present among the transferred cells, does antibody production commence without the injection of further antigen [31]. Then when increasing doses of the conjugate originally used for priming are given the antibody response first rises and then plateaus. The transferred cells are extremely sensitive to stimulation with certain antigens. With CGG, which is the most potent antigen encountered here, the cells detect as little as 0.001 μg of conjugate and reach a plateau at 0.1 μg. The response, which may extend over a ten thousandfold dose range, eventually declines with doses of over approximately 1 mg; this decline has been attributed to the induction of high-zone immunological tolerance [32].

Conjugates of the hapten with a protein other than that originally used for priming invariably elicit a response if given in a sufficient dose (Fig. 1). The cells are very much less sensitive to stimulation by these heterologous conjugates, although the slope of their dose-response curve is similar to that obtained with the homologous conjugate. Usually the response does not reach the plateau obtained with the homologous conjugate, but this can be attributed to entry into the zone of high-dose tolerance before the plateau can be attained.

Under these circumstances it would not be appropriate to compare the homologous and heterologous conjugates in respect of the amount of antibody which they can elicit, since this ratio will vary with the dose of antigen selected for test and since the peak doses are suspected of being affected to different extents by high-dose tolerance. The comparison was therefore made as follows in terms of *relative potency*, *i. e.* of the relative quantity of antigen required to reach a given level of stimulation. First, the data from 21 transfer experiments, distributed among various different conjugates as shown in Table 1, were selected.

Each experiment covered a wide enough dose range to contribute to the construction of a common curve. Those relating to one type of homologous conjugate were collected together and adjusted to a common plateau or 100 % value. In the case of CGG, OA, BGG and MGG, the mean of the respon-

41

Table 1. Levels of anti-NIP antibody attained after optimum stimulation of primed, transferred spleen cells by homologous conjugate as a function of interval after priming and carrier protein.

	Interval (weeks) from priming to cell transfer		
	5–10	10–20	20–30
Carrier protein		Anti-NIP	
	(l. u.)	(l. u.)	(l. u.)
BSA	1.46	0.50	
	0.95	1.05	
		1.58	
		0.78	
OA	1.72	1.42	1.82
	0.82	1.46	
		0.85	
CGG	1.34	1.85	
	1.20	2.35	
	2.26	2.00	
BGG	0.92		
	1.24		
MGG	1.70		

ses to 1 and 10 μg homologous conjugate were taken as the plateau and in the case of BSA, the mean of the 100 and 1000 μg responses. Fig. 1 illustrates the type of plot. Next, a common line was drawn by eye through all the data on the ascending part of the dose response curve, while permitting the intercept to vary according to which homologous conjugate had been used. This yielded a common slope of log response = 1.4 x log antigen dose. A good fit was obtained as illustrated in Fig. 1, even though the curve is more accurately sigmoid. (For this reason the fit was not formally computed). Each type of conjugate yielded its own intercept with the plateau, as will be shown in Table 3. Most of the intercepts are different, implying that different threshold doses of antigen are required to obtain maximum stimulation. Finally, the entire curve, consisting of slope and plateau with a constant intercept for each homologous conjugate, was fitted by eye to the data obtained from a series of transfer experiments. Data obtained with the variant haptens, e. g. NIP-ala$_4$, were fitted by means of the curve obtained for the corresponding NIP conjugate of the carrier protein.

A line with the same slope was drawn for the data for each heterologous conjugate in each experiment. The distance between the slopes for homologous and heterologous slopes

42

then yields the relative potency as shown in Fig. 2. The effect of this procedure is to obviate the variation which was encountered in the plateaux of response obtained with different cell populations, such as is shown in Table 1.

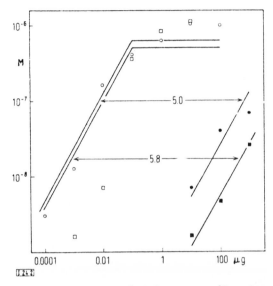

Figure 2. Measurement of relative potency of homologous and heterologous conjugates. Anti-NIP response (serum molar NIP-binding sites) as a function of dose of antigen (μg/mouse). Homologous conjugate (NIP-OA, open symbols), heterologous conjugate (NIP-BSA, closed symbols); two experiments, one marked as circles, the other as squares.

The values for relative potency obtained by this procedure are given in Table 2. In 30 out of 31 instances the heterologous conjugate proved less potent than the homologous, by factors varying up to a million-fold; a marked carrier effect was therefore obtained. It is also clear that the magnitude of the relative difference in potency varies for different pairs of conjugates. Rough averaging between reciprocal pairs, *e. g.* between secondary stimulation by NIP-OA of cells primed with NIP-BSA and by NIP-BSA of cells primed with NIP-OA, yields a 1000-fold difference between the potencies of the homologous and heterologous carriers. The variation in difference can be attributed to variation in 'intrinsic' potency of the conjugates, which is discussed below.

43

Table 2. Relative potency ($\log_{10}$) of NIP-carriers in inducing the secondary response (2).

		(1)				
		BSA	OA	CGG	BGG	MGG
(2)	BSA	0	-4.3[d] -4.4 -4.1 -4.6 -4.8[a] -5.6[b]	-4.4 -4.1 -5.1 -5.0[a] -5.8[c]		-2.6
	OA	-2.6 -2.6 -1.7[a] -0.0[b]	0	-6.0 -2.8[a]	-4.1	-3.4
	CGG	-1.9 -0.9 -1.2 +0.4[a]	-2.9 -2.1[a]	0	-2.7 -3.3 -3.3[c]	-1.7
	BGG			-5.1	0	

a) NIP-N^{11} (see section 2.3)
b) NIP-ala$_n$
c) NIP-ala$_4$
d) *i. e.* for cells obtained from mice immunized with NIP-OA in this experiment, NIP-OA is x $10^{4.3}$ as potent as NIP-BSA in stimulating the secondary response.

Table 2 includes data obtained with conjugates in which the NIP group is separated by various chemical spacer groups from the protein carrier. The local environment theory predicts that in these instances the carrier effect should be reduced, although the prediction is qualified by the possibility that flexible spacer groups may permit the hapten to fold over and remain in the hydrophobic local environment of the carrier [16]. The prediction is therefore strongest in the case of the relatively inflexible and less hydrophobic tetra-alanine spacer. No systematic reduction in the relative difference in potency could be detected with the use of spacers.

3.2. The 'intrinsic' potency of carrier protein

Conjugates of the same hapten prepared with different carrier proteins vary markedly in their capacity to elicit an anti-hapten response, as is well known. Data relevant to this variation are collected together in Table 3. The first column in the table refers to the intercept of the rising portion of the dose-response curve with the plateau, obtained by the procedure

which has just been described for the homologous carrier.
The second column refers to the corresponding intercept
for the heterologous carrier and was obtained by adjusting
the figure in the first column by the mean relative potency
taken from the data in Table 2. In making this adjustment
the mean was taken over all the conjugates used for priming;
the type of heterologous conjugate used for priming appeared
not to affect the potency of a conjugate in eliciting a secondary response. Note that the figures given in this column are
fictional, in the sense that 100 % stimulation cannot usually
be obtained with a heterologous conjugate, probably because
of the tolerance effect described above. The third column
gives the means of the data in Table 1.

An overall consistency can be seen in these data. Cells primed
with BSA can produce least antibody and BSA is a poor
carrier whether acting in a homologous or heterologous role.
The reverse is true of CGG. OA is an exceptionally poor heterologous carrier but is otherwise intermediate, a finding
which could be taken to imply that this carrier is exceptionally dependent on cell cooperation in induction of the response [33].

Table 3. Intrinsic potency of carrier proteins in the anti-NIP adoptive
secondary response

| Carrier proteins | Dose of antigen required for 100 % stimulation (μg) | | Plateau response |
	Homologous carrier	Heterologous carrier	Anti-NIP (l. u.)
BSA	100	800	1.04
OA	0.25	1500	1.34
CGG	0.1	30	1.83
BGG	0.1	–	1.08
MGG	1.0	–	1.18

The data obtained with the variant haptens clearly indicated
that the nature of the carrier, rather than the nature of the
hapten, played the dominant role in determining intrinsic
potency in the sense defined here. In fact no systematic difference could be found between the NIP-N[11], NIP-ala$_n$ and
NIP-ala$_4$ derivatives of any of the proteins tested.

3.3. Other variables

Over the interval examined no effect could be detected of varying the time after priming on the amount of antibody produced in the adoptive secondary response (see Table 1). Nor did the carrier effect diminish with time, contrary to other findings [34]. For example, the relative potency of BSA as a heterologous carrier for NIP-OA-primed cells was −4.4 log units in one transfer performed 6 weeks after priming, and −4.6 in another performed after 26 weeks.

Only minor effects could be obtained by varying the conjugation ratio. $NIP_{0.5}BSA$ and $NIP_{1.5}BSA$ primed less efficiently than $NIP_{11}BSA$, but $NIP_{3.3}OA$ primed as effectively as $NIP_{13}OA$ and $NIP_{2.6}CGG$ primed as effectively as $NIP_{6.6}CGG$. $NIP_{48}OA$ proved slightly more potent than $NIP_{13}OA$ in stimulating a secondary response when applied *in vitro*. Heterologous conjugates with high conjugation ratios, e. g. $NIP_{87}BGG$, were not notably more potent than the normal conjugates (5−15 moles/mole), nor were low ratios ($NIP_{1.5}$ BSA, $NIP_{1.4}OA$, $NIP_{2.6}CGG_1$) less potent.

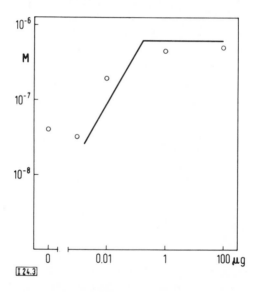

Figure 3. Secondary responses without cell transfer. Response of $NIP_{13}OA$-primed mice (serum molar NIP-binding sites) as a function of dose of antigen (μg $NIP_{13}OA$/mouse).

46

Cells exposed to antigen *in vitro* showed the same carrier effect as those exposed after transfer. In one such experiment NIP-BSA gave a relative potency of -4.5 log units compared with the homologous conjugate, NIP-OA. In another NIP-BSA gave a relative potency of -4.7 log units compared with the homologous conjugate, NIP-CGG.

The secondary response as studied here displays much the same qualitative features as the secondary response obtained in the conventional manner by boosting without cell transfer. This is illustrated in Fig. 3. Note that in the experiment shown in this figure which was performed on mice primed 22 weeks earlier, both the threshold of response and the total amount of antibody produced falls within the range obtained above with the transfers. The non-transfer experimental design is open to two objections: one is that some background antibody is present and the other that individual mice may have been primed with varying efficiency.

3.4. Lack of a primary response to the 'secondary' stimulation

The possibility arises that the 'secondary' response obtained with large doses of heterologous conjugate might in fact be due to a primary response. The responses were accordingly examined of irradiated mice which received a transfer of either normal cells or of cells primed with a protein other than that in the conjugate. With neither NIP-OA nor NIP-BSA in 1 mg doses could any response be detected. With NIP-CGG, in this one dose, a response of questionable significance ($10^{-7.3}$ M hapten binding capacity) was obtained. This would not have contributed significantly to a measurement of relative potency, so that the potential threat can be dismissed.

3.5. Antigen contamination

The danger of contamination when one antigen is used in a dose one million-fold higher than another is appreciable. Contamination was monitored from time to time throughout the series of experiments by a method used before [25] in which antibody to the homologous carrier protein (*e. g.* CGG) was assayed in those sera containing anti-hapten antibody elicited by the 'heterologous' conjugate (*e. g.* NIP-BSA). No such antibody could be detected and so contamination had apparently been avoided, except as expected in the case of BGG which was contaminated with fairly large amounts of BSA ('relative potencies' in this connection are of course omitted from Table 2).

47

3.6. Inhibition by excess carrier protein

Inhibition by carrier protein represents another kind of experiment in which the predictions made by the local environment theory break down. The anti-hapten adoptive secondary response to NIP_7CGG can be inhibited by the carrier CGG, although the anti-hapten antibody has no detectable affinity for the carrier [17,26]. A similar but less dramatic inhibition can be demonstrated with OA and BSA as carriers. This is illustrated in Fig. 4, which includes both an *in vivo* and an *in vitro* stimulation experiment. As before, the inhibition is competitive and can be largely overcome by increasing the dose or concentration of the conjugated antigen. The inhibition obtained with OA is relatively inefficient, a finding which suggests that the carrier determinants present on the $NIP_{11}OA$ conjugate are not all represented on unconjugated OA [33].

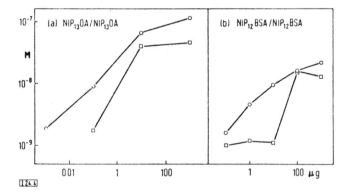

Figure 4. Inhibition of adoptive secondary response by excess carrier protein. Anti-NIP response (serum molar NIP-binding sites) as a function of dose of conjugate (μg/mouse). (a) Cells from $NIP_{13}OA$-primed donors stimulated with $NIP_{13}OA$; circles without, squares with 1000 μg OA. *In vivo* stimulation. (b) Cells from $NIP_{12}BSA$-primed donors stimulated with $NIP_{12}BSA$; circles without, squares with 10 000 μg BSA. *In vitro* stimulation.

The lack of affinity for carrier on the part of hapten-binding antibody has been verified with OA and BSA. The hapten binding capacity of antisera raised to $NIP_{11}OA$ and NIP_{14} BSA was measured in the presence and absence of OA and BSA respectively, each at concentrations of $N^{131}IP$-CAP 10^{-8} M and protein 1 mg/ml: no inhibition of binding by the carrier protein could be detected.

These findings of carrier-inhibition cannot easily be reconciled with the local environment hypothesis. It might be argued that minor contributions to the energy of binding may be of great importance in the stimulation of cells and that the serological test simply failed to detect these contributions. On the other hand the concentration of protein used here for inhibition is, if anything, less than the concentrations of hapten hitherto used for inhibition of stimulation, while the affinity for antibody on the part of the hapten is certainly very much larger.

4. Discussion

A consistent and powerful carrier effect has been obtained with all the carrier proteins tested — BSA, OA, CGG and BGG — in conjugates with NIP and its structurally related haptens. This confirms the earlier work cited above and extends it in the sense that measurements were made of the relative potency of carrier proteins as homologous and heterologous carrier. These measurements of relative potency were made possible by the consistent ability of heterologous conjugates to elicit a detectable response if administered in sufficiently high doses. The ability of high doses of antigen to override the carrier effect has been detected in one previous study [10] but not in others [8, 35]. Previous failures to override can be attributed to technical considerations, such as the use of relatively small doses of antigen on a body weight basis ($<$ 10 mg in rabbit) or less sensitive methods of detecting anti-hapten antibodies. Alternatively, they may reflect genuine differences in the way other species handle other conjugates.

The ability of heterologous conjugates to induce has another interest, for it enables the dose response relationship to be compared for homologous and heterologous carrier. The fact that they show parallel slopes immediately suggests that the response to the carrier acts by magnifying the effective dose of antigen, and indeed, it was on these grounds that the local concentration hypothesis was originally formulated [6].

Another example of overriding the carrier effect can be found in the ability of moderate doses of DNP-KLH conjugates to elicit an anti-DNP response in rabbits primed with DNP-BGG [36, 37]. An explanation of the exceptional case has been sought in the possibility that such a powerful immunogen as DNP-KLH may elicit a primary response against its own carrier determinant [38]. Alternatively, KLH may simply have

49

sufficient intrinsic potency as an immunogen to render the carrier effect redundant. In an analogous series of experiments performed on rabbits primed with TNP-KLH, the absence of carrier effect can perhaps be attributed to the use of larger doses (15 mg) of antigen in the boost [39].

The nature of intrinsic immunogenic potency merits further attention. It reflects, presumably, the ability of an antigen to exploit local concentrating mechanisms other than the carrier effect. A provisional listing and classification of other known mechanisms has been drawn up [40]. At present the list includes multivalent binding to receptors, electrostatic attraction to receptors, binding to macrophages and 19 S antibody. The contribution of these various mechanisms to the observed difference between CGG and BSA, to take extreme examples from the present data, cannot yet be properly assessed.

The figures presented in Table 3 underline the need to examine doses of antigen over a wide range. Had a narrow series of doses been chosen, in the range 50–500 μg antigen, NIP-BSA would have been found consistently carrier-dependent and NIP-CGG consistently independent. In terms of the thymus-derived helper cell hypothesis [33] this would be equivalent to finding the anti-NIP response thymus-dependent in one case and thymus-independent in the other.

The figures in Table 3 specify the optimum combination of carrier protein for further use in the analysis of helper cells. OA or CGG are the best carriers among those tested for priming mice for an anti-hapten response. BSA is a good heterologous carrier for eliciting a secondary response because of its low intrinsic potency and consequent sensitivity to the presence of helper cells.

I thank Messrs Burroughs Wellcome for a gift of the Bordetella pertussis, Miss Laura Tackett for writing the computer programs, Dr. Paul Plotz for developing the antiimmunoglobulin co-precipitation test and Miss Doreen Cross, Miss Margo Peacock and Miss Annabel Green for skilled technical assistance.

Received September 6, 1970.

5. References

1 Claman, H. N., Chaperon, E. A. and Triplett, E. L., *Proc. Soc. Exp. Biol. Med.*, 1966. *122*: 1167.

2 Davies, A. J. S., *Transplantation Reviews* 1969. *1*: 43.

3 Miller, J. F. A. P. and Mitchell, G F., *Transplantation Reviews* 1969. *1*: 3.

4 Taylor, R. B., *Transplantation Reviews* 1969. *1*: 114.

5 Mitchison, N. A., *Progr. Biophys.* 1966. *16*: 3.

6 Mitchison, N. A., *Differentiation and Immunology* in Warren, K. B. (Ed.) *Symp. Int. Cell. Biol.* 1968. 7: 29.

7 Salvin, S. B. and Smith, R. F., *Proc. Soc. Exp. Biol. Med.* 1960. *104*: 584.

8 Ovary, Z. and Benacerraf, B., *Proc. Soc. Exp. Biol. Med.* 1963. *114*: 72.

9 Weigle, W. O., *J. Exp. Med.* 1962. *116*: 913

10 Benacerraf, B. and Gell, P. G. H., *Immunology* 1959. 2: 53.

11 Rajewsky, K. and Rottländer, E., *Cold Spring Harbor Symp. Quant. Biol.* 1967. *32*: 547.

12 Plescia, O. J., Palczuk, N. C., Braun, W. and Cora-Figueroa, E., *Science* 1965. *148*: 1102.

13 McBride, R. A. and Schierman, L. W., *Science* 1966. *154*: 655.

14 Levine, B. B. *J. Exp. Med.* 1965. *121*: 873.

15 Haber, E., Richards, F. F., Spragg, J., Austen, K. F., Vallotton, M. and Page, L. B., *Cold Spring Harbor Symp. Quant. Biol.* 1967. *32*: 299.

16 Singer, S. J., *Immunochemistry* 1964. *1*: 15.

17 Mitchison, N. A., *Cold Spring Harbor Symp. Quant. Biol.* 1967. *32*: 431.

18 Rajewsky, K., Rottländer, E., Peltre, G. and Müller, B., *J. Exp. Med.* 1967. *126*: 581.

19 Bretscher, P. A. and Cohn, M., *Nature* 1968. *220*: 444.

20 Mitchison, N. A. in Mitchison, N. A. Greep, J. M. and Hattinga Vershure J. C. M. (Eds.) *Organ Transplantation Today.* Excerpta Medica Foundation, Amsterdam 1969. p. 13.

21 Mitchison, N. A. in *Immunological Tolerance.* Academic Press, N. Y. 1969, p. 113

22 Boak, J. L., Kölsch, E. and Mitchison, N. A., *Antibiot. Chemother.* 1969, *15*: 98

23 Mitchison, N. A., *Transplant. Proc.* 1970. 2: 92.

24 Dresser, D. W., *Immunology* 1965. 9: 261.

25 Brownstone, A., Mitchison, N. A. and Pitt-Rivers, R., *Immunology* 1966. *10*: 481.

26 Brownstone, A., Mitchison, N. A. and Pitt-Rivers, R., *Immunology* 1966. *10*: 465.

27 Pitt-Rivers, R., Meyer-Delius, M., Rüde, E. and Mitchison, N. A. – in Preparation.

28 Axen, R., Porath, J. and Ernback, S., *Nature* 1967. *214*: 1302.

29 Mitchison, N. A., *Proc. Roy. Soc. Lond.* B. 1964. *161*: 275.

30 Mitchison, N. A. – unpublished data.

31 Mitchison, N. A., *Israel J. Med. Sci.* 1969. 5: 230.

32 Mäkelä, O. and Mitchison, N. A., *Immunology* 1965. *8*: 549.

33 Mitchison, N. A., *Eur. J. Immunol.* 1971. *1*:18.

34 Paul, W. E., Siskind, G. W., Benacerraf, B. and Ovary, Z.
   J. Immunol. 1967. *99*: 760.

35 Rajewsky, K., Schirrmacher, V., Nase, S. and Jerne, N. K.,
   J. Exp. Med. 1969. *129*: 1131.

36 Paul, W. E., Siskind, G. W. and Benacerraf, B., *Immunology*
   1967. *13*:147.

37 Eisen, H. N., Little, J. R., Steiner, L. A., Simms, E. S. and Gray, W.,
   Israel J. Med. Sci. 1969. *5*: 338.

38 Taylor, R. B. and Iverson, G. M., *Proc. Roy. Soc. Lond.* B. 1970.
   in press.

39 Rittenberg, M. R. and Campbell, D. M., *J. Exp. Med.* 1968. *127:* 717.

40 WHO *Techn. Res. Ser.* 1970. *No. 448,* p. 49.

Cooperation between Carrier-reactive and Hapten-sensitive Cells *in vitro*

Christina Cheers
J. C. S. Breitner
Margery Little
J. F. A. P. Miller

THE mechanism, known as the carrier effect, whereby immunity to one or more determinant groups enhances the response to other determinants on the same multivalent antigen, was first recognized in delayed hypersensitivity to haptens, in which, for an appreciable response, the hapten must be coupled to the same protein carrier for priming and challenge[1,2]. Carrier specificity has also been demonstrated in the secondary antibody responses to hapten protein conjugates[3]. Two alternative hypotheses have been advanced to explain this specificity. The "local environment" hypothesis supposes that the hapten-sensitive cell recognizes both the hapten and the carrier determinants. However, the antihapten antibodies produced do not distinguish details of the carrier molecule and so do not reflect the specificity of the cellular receptoi. Furthermore, inert spacer molecules inserted between hapten and carrier do not interfere with carrier specificity in the antibody response[3]. Reflecting current views on the cooperation between thymus-derived (T) and bone marrow derived (B) lymphocytes in the antibody response to various antigens[4], the second hypothesis invokes two or more cells, one with receptors directed towards the hapten (hapten-sensitive cell), the others specific for the carrier molecule proper (carrier-reactive cells). Supporting this is the observation that pre-immunization to a particular protein carrier alone could potentiate the primary or secondary antihapten response to a hapten conjugated to that protein[5]. In an adoptive transfer system, moreover, the efficiency of antihapten antibody production by cells primed to a particular hapten–protein conjugate and stimulated with the hapten conjugated on a heterologous protein, is significantly enhanced by the introduction of cells primed to the heterologous carrier alone. Anti-carrier serum antibody does not cause such enhancement[6]. The carrier-reactive cells must therefore cooperate in increasing the efficiency of the hapten-sensitive cells in some way other than by providing humoral anti-carrier antibody. Recent work strongly suggests that carrier reactive cells are thymus-derived[6,7].

Production of the carrier effect *in vitro* would enable further studies of the mechanism of cell to cell interaction in antibody responses in a system that lacks some of the complexities of the *in vivo* situation. For example, it would allow the identification of the antibody forming cell. It would also avoid diversion of cells pretreated in various ways to sites where they are no longer able to function. Lesley and Dutton[8] have reported inhibition of the antibody response to sheep erythrocytes in mouse spleen cells cultured in the continuous presence of anti-K chain serum, or pretreated with serum and complement, an experiment which *in vivo* would have been complicated by opsonization of the cells. We have therefore investigated the antibody response of mouse spleen cells to hapten–protein conjugates *in vitro*. A preliminary report of this work has already been given[9].

The hapten, 3,5-dinitro,4-hydroxyphenylacetic acid (NNP) was coupled to the non-cross reacting proteins, fowl immunoglobulin G (FγG) and ovalbumin (OV), by the method of Brownstone et al.[10]. There were between five and ten NNP groups per molecule of protein. The mice, usually CBA, were immunized at 6–8 weeks of age with 800 µg of protein antigen or hapten–protein conjugate precipitated on alum and injected intraperitoneally together with 2×10^9 *Bordetella pertussis* organisms. After 4–8 weeks, spleens from immunized or normal mice of the same age were teased through an 80-mesh stainless steel sieve. A total of 3×10^6 cells in 1 ml. of medium was placed in the inner culture chamber of a Marbrook–Diener type flask[11,12], together with 10 µg of fluid antigen, an amount found to be optimal in preliminary experiments. Eagle's minimal essential medium with 5% foetal calf serum, 100 mg streptomycin sulphate and 60 mg penicillin G per litre was used and the cultures were incubated at 37° C in 10% CO_2 in air for 4 days. The cells were then collected and assayed for antibody-forming cells using the Cunningham technique[13]. The number of NNP-specific plaques was determined by subtracting the number of lytic plaques obtained with unconjugated sheep erythrocytes from that obtained with NNP-conjugated sheep erythrocytes. Conjugation of NNP to sheep erythrocytes was performed according to the method of Pasanen and Makela[14]. When mixed populations of cells from CBA and (CBA × C57BL)F$_1$ mice were used, the identity of the plaque forming cells could

54

be determined using plaque inhibition by specific allo-antisera[4].

Normal spleen cells exposed *in vitro* to NNP-protein conjugate did not respond. Primed spleen cells gave a response of the order of 5–20 times higher than the sheep cell background. In most experiments, NNP.FγG was used as the antigen *in vitro* because of the weaker immunogenicity of NNP.OV. Variation in the peak response from one experiment to another made it impossible to combine results, so use was made of the relationship demonstrated by Fisher[15] for accumulating the probabilities of separate experiments.

Table 1 demonstrates the carrier specificity of the *in vitro* response. The protein used for priming the mice *in vivo* and for stimulating the cells *in vitro* had to be identical (homologous) in order to obtain the optimal yield of NNP-specific

Table 3 Inhibition of Carrier Cooperation by Anti-immunoglobulin

Antigen *in vivo*	Cells pretreated	Antigen *in vitro*	Anti-NNP PFC per culture§
NNP.OV * FγG *	— —	NNP.FγG	595, 375
NNP.OV * FγG *	— NRS†	NNP.FγG	1,046, —
NNP.OV * FγG *	— Anti-immunoglobulin‡	NNP.FγG	61, 0
NNP.OV * FγG *	— —	NNP.FγG	136, 375
NNP.OV * FγG *	NRS† —	NNP.FγG	218, —
NNP.OV * FγG *	Anti-immunoglobulin‡ —	NNP.FγG	59, 75

* Cells from different donors.
† NRS, normal rabbit serum.
‡ Rabbit anti-mouse polyvalent immunoglobulin.
§, Geometric mean of four cultures in each of four experiments.

plaques. Since this was observed when either NNP.FγG or NNP.OV were used as priming or challenge antigen, the difference in the results cannot be ascribed to the intrinsically stronger immunogenicity of NNP.FγG. The NNP plaque-forming cell response stimulated by heterologous carrier-NNP conjugate could be enhanced significantly by adding to the culture system spleen cells primed only to the heterologous carrier (Table 2). Carrier primed cells alone did not produce antihapten plaques when cultured with NNP-protein conjugates, although they did so when erythrocytes were the carrier (ref. 16 and unpublished observations).

It has been postulated that the cell receptor is an immunoglobulin molecule[17-19]. Spleen cell suspensions were incubated at 4° C for 3 h in tissue culture medium containing 20% rabbit antimouse polyvalent immunoglobulin serum or 20% normal rabbit serum. The cells were then washed three times and cultured in the usual way. As shown in Table 3, pre-

Table 1 Carrier Specificity *in vitro*

Antigen *in vivo*	Antigen *in vitro*	Anti-NNP p.f.c. per culture*									Accumulated probability†
NNP.FγG	NNP.FγG	332	859	103	208	603	175	150	23	326	<0.01
NNP.OV	NNP.FγG	225	22	101	2	25	34	104	11	209	
P value for individual experiments		0.25	0.02	0.51	0.05	0.13	0.13	0.25	0.19	0.19	
NNP.OV	NNP.OV	201	206	12	48	130	433	89	156	159	<0.01
NNP.OV	NNP.FγG	225	22	87	25	34	104	200	27	51	
P value for individual experiments		0.72	0.01	0.86	0.41	0.06	0.01	0.99	0.02	0.02	
None	NNP.FγG	7	36	23	100	14					

* Each number represents the geometric mean of three or four cultures and is paired with the contrasting result from the same experiment.
† Where each of n experiments yields a probability P, $-2 \log_e P$ is distributed as χ^2 with $2n$ degrees of freedom[15] and an accumulated measure of significance can be derived.

Table 2 Carrier Cooperation *in vitro*—Effect of Syngeneic Spleen Cells Primed to Carrier

Antigen *in vivo*	Antigen *in vitro*	Anti-NNP PFC per culture*									Accumulated probability*
NNP.FγG†	NNP.FγG	332	859	103	208	603	175	150	23	326	<0.01
NNP.OV†	NNP.FγG	225	22	101	2	25	34	104	11	209	
NNP.OV + FγG‡	NNP.FγG	549	745	685	86	247	321	491	120	290	<0.001
P values for differences between results in lines 2 and 3 above in individual experiments		0.02	0.02	0.05	0.05	0.19	0.05	0.02	0.02	0.18	

* See footnotes to Table 1. † Supplemented with normal spleen cells. ‡ Two separate donor sources.

Table 4 Carrier Cooperation Between Semi-allogeneic Combinations

Antigen *in vivo*	Mouse strain	Antigen *in vitro*	Anti-NNP PFC per culture	Reduction by CBA	C57
NNP.FγG None	CBA F₁*	NNP.FγG	186	97%	24%
NNP.OV None	CBA F₁	NNP.FγG	34	—	—
NNP.OV FγG	CBA F₁	NNP.FγG	338	94%	0%
NNP.FγG None	F₁ CBA	NNP.FγG	208	83%	86%
NNP.OV None	F₁ CBA	NNP.FγG	23	—	—
NNP.OV FγG	F₁ CBA	NNP.FγG	185	76%	84%

* (CBA × C57Bl)F₁ hybrids.

treatment either of the carrier primed cells or of the hapten sensitive cells with anti-immunoglobulin serum produced a marked inhibition of the production of anti-NNP plaque-forming cells. Semi-allogeneic combinations of cells were able to cooperate *in vitro* as effectively as syngeneic cells (Table 4). Most of the anti-NNP plaque forming cells in these cultures were derived from the hapten-primed cell population, not from the carrier-primed cells.

To test the helper function of a population of thymus derived cells, CBA mice were subjected to 750 rads total body irradiation and injected intravenously with 100 million syngeneic thymus cells and intraperitoneally with alum-precipitated protein. No *B. pertussis* was given nor did the mice receive bone marrow cells. The spleens of these mice were collected after 5–7 days and cell suspensions were made and cleaned of debris by layering twice on foetal calf serum and allowing them to settle for 10 min. As few as 5×10^6 of these "activated thymus cells" substituted effectively for carrier-primed, whole spleen cells in enhancing the response of cells primed to a heterologous NNP-protein conjugate (Table 5). The effect was specific since thymus cells activated to FγG enhanced the response induced by NNP.FγG while those activated to bovine serum albumin did not.

The results presented here are in general agreement with those obtained by others using *in vivo* systems[6]. We have demonstrated that the carrier effect can be reproduced *in vitro* and that it is specific. In addition, the evidence strongly supports the notion that separate cells are carrier-reactive and hapten-sensitive. Either population could be inhibited by treatment before culture with anti-immunoglobulin serum, implying possible interference with the antigen receptor site

Table 5 Activated Thymus Cells as Carrier Primed Cells

Antigen *in vivo*	Antigen *in vitro*	Anti-NNP PFC per culture
NNP.FγG None	NNP.FγG	603, 181
NNP.OV None	NNP.FγG	25, 59
NNP.OV FγG	NNP.FγG	247, 268
NNP.OV FγG *	NNP.FγG	480, 160
NNP.OV BSA *	NNP.FγG	— 69

See footnote to Table 1.
In each case the source of cells was the spleen.
* Mice receiving 750 rad X-irradiation were injected intravenously with 100×10^6 thymus cells from young syngeneic donors, and intraperitoneally with 800 μg alum precipitated protein. Between 5 and 7 days later their spleens were used to prepare suspensions of "activated thymus cells" (ATC).

on the cells. The carrier reactive population was not directly concerned with formation of antibody to the hapten. This occurred preferentially among cells previously primed to the hapten. Thymus derived cells acted as carrier-primed cells and were able to cooperate with spleen cells from mice primed to hapten conjugated onto a heterologous carrier. Their role was specific, since thymus derived cells primed to bovine serum albumin did not enhance the response to NNP.FγG. This finding confirms and extends the original observation that the ability of thymus-derived cells to cooperate with antibody forming cell precursors could be specifically enhanced by prior *in vivo* exposure to the relevant antigen[20].

This work was supported by the National Health and Medical Research Council of Australia, the Australian Research Grants Committee, the Damon Runyon Memorial Fund for Cancer Research, the British Heart Foundation and the National Heart Foundation of Australia. Anti-immunoglobulin serum was supplied by Dr N. L. Warner.

[1] Benacerraf, B., and Gell, P. G. H., *Immunology*, **2**, 53 (1959).
[2] Benacerraf, B., and Gell, P. G. H., *Immunology*, **2**, 219 (1959).
[3] Mitchison, N. A., *Cold Spring Harbor Symp. Quant. Biol.*, **32**, 431 (1967).
[4] Miller, J. F. A. P., and Mitchell, G. F., *J. Exp. Med.*, **128**, 801 (1968).
[5] Katz, D. H., Paul, W. E., Goidl, E. A., and Benacerraf, B., *J. Exp. Med.*, **132**, 261 (1970).
[6] Mitchison, N. A., *Europ. J. Immunol.*, **1** (in the press).

[7] Miller, J. F. A. P., *Proc. Third Sigrid Juselius Symp.*, Helsinki (1970).
[8] Lesley, J., and Dutton, R. W., *Science*, **169**, 487 (1970).
[9] Breitner, J., and Miller, J. F. A. P., *Fed. Proc.*, **29**, 572 (1970).
[10] Brownstone, A., Mitchison, N. A., and Pitt-Rivers, R., *Immunology*, **10**, 465 (1966).
[11] Marbrook, J., *Lancet*, ii, 1279 (1967).
[12] Robinson, W. R., Marbrook, J., and Diener, E., *J. Exp. Med.*, **126**, 347 (1967).
[13] Cunningham, A. J., and Szenberg, A., *Immunology*, **14**, 399 (1968).
[14] Pasanen, V. J., and Makela, O., *Immunology*, **16**, 399 (1969).
[15] Fisher, R. A., *Statistical Methods for Research Workers*, 99 (Oliver and Boyd, Edinburgh, 1954).
[16] Trowbridge, I. S., Lennox, E. S., and Porter, R. R., *Nature*, **228**, 1087 (1970).
[17] Makela, O., *Transplant. Rev.*, **5**, 3 (1970).
[18] Sell, S., *Transplant. Rev.*, **5**, 19 (1970).
[19] Greaves, M. F., *Transplant. Rev.*, **5**, 45 (1970).
[20] Mitchell, G. F., and Miller, J. F. A. P., *Proc. US Nat. Acad. Sci.*, **59**, 296 (1968).

DYNAMIC ASPECTS OF ANTIGEN BINDING TO B

CELLS IN IMMUNE INDUCTION

J.D. Wilson & Marc Feldmann

Antigens either immunize B lymphocytes direct, or must be aided by such cells as T cells or macrophages. This is chiefly caused by the physical structure of the antigen's macromolecular "carrier".

COLLABORATION between thymus-derived (T) and thymus-independent (B) lymphocytes is an essential step in the antibody response to many[1-5] but not all antigens[6], and certain rules have recently been defined to tell whether antigens may immunize B cells directly or only through the assistance of helper cells, such as T cells and macrophages[7]. Large polymeric antigens with repeating antigenic determinants, such as polymeric flagellin of *Salmonella adelaide* (POL) or the dinitrophenyl group (DNP) conjugated to POL (DNP, POL), induce antibody responses in the absence of T cells[6], carrier reactive cells, or macrophages[7]. By contrast, antibody responses to the same antigenic determinants (DNP) conjugated to macromolecules lacking repeating units, such as monomeric flagellin (MON) or gamma globulin, could not immunize in the absence of helper effect of both T cells[6] and of macrophages (unpublished work of M. F.). Thus depending on the physical structure of its macromolecular "carrier" an antigenic determinant such as DNP may interact with the immunoglobulin (Ig) receptors for antigen on B cells[8] in an immunogenic or a nonimmunogenic way. Recent experiments (unpublished work of J. D. W., G. J. V. Nossal and H. Lewis) using radioiodinated rabbit antimouse Ig sera have demonstrated that surface immunoglobulin is rapidly lost from the cell surface and is replaced. By direct radio labelling of the cell surface, Cone et al.[9] have confirmed that Ig is lost from the cell surface. These experiments suggested an explanation for the different cellular pathways of immune induction of an anti-DNP response with DNP POL (or DNP Flagella) and DNP HGG.

The binding of antigen to B cell receptors was demonstrated by the rosette formation to DNP-coated sheep red blood cells (DNP-SRBC). The loss of antigen from the cell surface, with the subsequent reappearance of specific receptors, was monitored by rosette formation in mouse spleen cell population which were initially blocked with saturating concentrations of DNP on various carriers. It was established that DNP on a carrier with repeating units (DNP-Fla), remained essentially fixed to the cell surface receptors at 37° C, whereas DNP on smaller molecules (DNP lysine, or DNP_{12} HGG) was rapidly lost from the lymphocyte surface, exposing fresh receptors.

CBA or $(CBA \times C57Bl)F_1$ mice were injected intraperitoneally with 100 μg of DNP fowl γ-globulin (DNP FγG) or 100 μg of DNP human γ-globulin (DNP HGG) emulsified in Freund's complete adjuvant, 6 and 2 weeks before they were killed. These spleens contained $1,950 \pm 320$ DNP specific rosettes 10^6 nucleated cells, and more than 90% of the rosette-forming cells (RFC) were resistant to treatment with AKR anti-θ C_3H serum and complement, and were thus B cells[11].

Spleens were removed and the cells washed three times in Eagle's minimum essential medium (Grand Island Biological Co.) buffered to pH 7.3 with HEPES (N-2-hydroxy ethylpiperazine N-1 ethanesulphonic acid, Calbiochem) ('HEM'), and resuspended at 10^7 nucleated cells per ml. in HEM containing 10% foetal calf serum (FCS) by volume. The cells were then incubated for 30 min at 4° C with saturating concentrations of DNP attached to various carriers: 10^{-4} M ϵ-2, 4-DNP-L-lysine (Mann Research Lab.), 7×10^{-7} M $DNP_{12}HGG$[12], and 8×10^{-8} M DNP flagella (DNP Fla)[12]. The substitution rate of DNP Fla is expressed as the number of DNP groups per unit of monomeric flagellin in the flagella. Thus there were three physical forms of the DNP determinant available to react

61

with receptors: monovalent DNP-lysine, multivalent DNP_{12} HGG, which, having the same diameter as a cell surface Ig receptor[13], could thus only form two or occasionally three bonds; and DNP Fla, a long molecule (up to 1 μm)[14], which exhibits many hundreds of antigenic determinants and so can make bonds with many receptor antibody molecules. Concentrations of DNP in these three forms needed to inhibit B rosette formation were determined as described elsewhere (unpublished work of J. D. W. and M. F.). Multiple samples were treated with each DNP compound. Matching control samples were incubated without DNP, but otherwise treated identically.

SRBC-coated with DNP-rabbit anti-SRBC-Fab'[15] were added to 1.0 ml. samples in plastic conical centrifuge tubes; the cells were syringed through a 26-gauge needle to break up clumps, then centrifuged for 6 min at 400g at 4° C. After a further 10 min the pellets were gently resuspended with a Pasteur pipette and rosettes were counted in a haemocytometer[16]. Only white cells completely surrounded with DNP-SRBC, with at least 8–10 SRBC, were classified as "rosettes". Before rosette preparation, multiple cell samples, both DNP-treated and untreated controls, were washed in HEM with 10% FCS and resuspended at 2×10^6/ml. in 5 ml. of HEM with 10% FCS and incubated at 37° C or 4° C. At intervals of up to 4½ h individual DNP-treated samples were removed together with their controls, centrifuged, resuspended at 10^7 spleen cells/ml. in phosphate buffered saline (PBS) with 10% FCS. DNP-SRBC were added and rosettes prepared as above. The percentage inhibition of rosette formation was calculated from the differing rosette numbers in the DNP-treated samples and their matched controls.

Six rosette inhibition experiments were performed. The results were pooled, standardized and are shown in Fig. 1. The patterns of recovery were different with each compound. (a) DNP-lysine: inhibition of rosette formation was rapidly reversed both at 4° C and at 37° C. Even after one wash at 4° C inhibition, the degree of the inhibition had diminished from 80% to less than 40%, and after a further hour at 4° C or at 37° C there was insignificant difference in rosette numbers between DNP-lysine-treated cells and their controls. (b) $DNP_{12}HGG$: spleen cells incubated with $DNP_{12}HGG$ at 4° C did not recover their capacity to form rosettes even after 4½ h at 4° C. In marked contrast, there was a rapid reversal of the inhibition of RFC after incubation at 37° C, which was almost as rapid as that with DNP-lysine. (c) DNP-Fla: inhibition was not reversed in washed cells incubated at 4° C. Inhibition of RFC was only slight, declining from 77% to 54% in 4½ h at 37° C. Thus the recovery of receptors on RFC treated with DNP-Fla, had a half-life of about 9 h.

Thus the rate of recovery of receptors on RFC varied

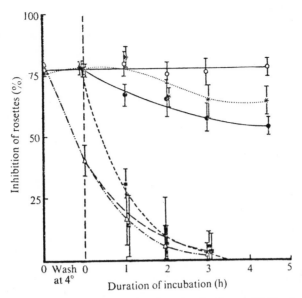

Fig. 1 Recovery of rosette-forming capacity from inhibition by saturating concentrations of DNP coupled to different carriers: the effects of incubation at $4°$ C and $37°$ C. Each point represents the arithmetic mean $\pm$ s.e. for three experiments. $*---*$, DNP HGG $37°$ C; $\triangle - \cdot \cdot - \triangle$, DNP-Lys $4°$ C; ■——■, DNP-Lys $37°$ C; $* \cdots *$, DNP-HGG $4°$ C; ○, DNP-Fla $4°$ C; ●, DNP-Fla $37°$ C.

greatly, depending on the potential number of bonds formed by the inhibitor with which they were pretreated. We then examined the antigens used to inhibit RFC for their capacity to induce an immune response *in vitro*, and so correlate the dynamic pattern of antigen binding with the process of immune induction in B cells. Since DNP-lysine is non-immunogenic *in vitro*, as *in vivo* the responses to pulses of $DNP_{12}HGG$, $DNP_{12}F\gamma G$ and DNP_1 Fla were evaluated. Culture methods, media, and antibody-forming cell (AFC) assays have been described in detail elsewhere[12,15]. Similar concentrations of $DNP_{12}HGG$, $DNP_{12}F\gamma G$ and $DNP_{1.5}Fla$, with the same degree of conjugation were used in both the *in vitro* immune response and rosette inhibition experiments. Spleen cells of DNP-primed CBA mice (see above) were treated with 10 μg/ml. $DNP_{12}HGG$, 10 μg/ml. $DNP_{12}F\gamma G$ and 10 μg/ml. DNP-Fla *in vitro* for periods of 15 min, 60 min, or 6 h at $4°$ C or $37°$ C before being thoroughly washed and cultured for 4 days with donkey red cells (DRBC) as a specificity control, but without any additional DNP conjugates. The combined results of 3 such experiments are shown in Fig. 2.

Pretreatment with DNP-Fla markedly stimulated the production of anti-DNP AFC, even if present for a very short period of time at $4°$ C or $37°$ C. Essentially complete stimulation was caused by 60 min exposure to an "obligatory immunogenic" preparation of DNP Fla[11]. In contrast pretreatments with neither DNP FγG nor DNP HGG caused any response. If the conjugate used for priming the mice $DNP_{12}HGG$ was used to stimulate the cells in culture, a response occurred, but only provided the antigen is left for more than 6 h.

The capacity to immunize B cells by this method correlates well with the binding of antigen to the cell surface. Thus DNP-lysine forming only a single bond with the receptor molecule uncouples rapidly from the cell surface receptors, since elution of DNP-lysine was virtually the same at 4° C as at 37° C as would be expected from thermodynamic considerations[18]. In contrast, $DNP_{12}HGG$ binds more firmly to the B cell surface, as one thousand times less $DNP_{12}HGG$ than DNP-lysing (or $DNP_{0.3}HGG$) inhibits the response to DNP POL *in vitro* (unpublished work of M. F.). This much greater efficiency of suppression indicates that $DNP_{12}HGG$ forms more than a single bond with the cell surface. These results are closely analogous to the much greater stability of binding of antibody molecules to multivalent $DNP\varphi X_{174}$ than monovalent DNP-lysine found by Hornick and Karush[19], and accounts for the insignificant decline in rosette inhibition over 4.5 h at 4° C. Similarly multivalent DNP-Fla did not permit the recovery of RFC at 4° C.

In contrast, cells left to recover at 37° C yielded different results. Rosette inhibition by $DNP_{12}HGG$ rapidly declined at 37° C, with a half-life of 1.25 h, whereas the inhibition by DNP-Fla was minimally reversible with a half-life of 9 h.

There are two possible interpretations of the lack of recovery of RFC. The simplest interpretation, and the one that holds for both 4° C and 37° C is that antigen is still bound to the surface receptors preventing rosette formation. Alternatively, the receptors for antigen may no longer be present on the cell surface. They may form caps, and become pinocytosed, as recently described by Taylor *et al.*[20] for mouse lymphocytes treated with antimouse Ig serum, and with multivalent DNP BSA but not with monovalent DNP BSA. Since $DNP_{12}HGG$ is multivalent it could have induced "cap" formation, pinocytosis and the loss of surface Ig at 37° C. However, rosette formation rapidly recovered in our experiments at 37° C suggesting more that loss of the Ig receptor-antigen complex into the medium had occurred, with subsequent rapid receptor replacement, in keeping with the results of Wilson *et al.* (to be published) who found that recovery of rosette formation at 37° C from anti-κ serum inhibition had an almost identical half-life. Conceivably different concentrations of antigen may induce pinocytosis or elution of surface receptors. We used 10 μg/ml. which was immunogenic *in vitro* (Fig. 2)[11]. The failure of receptors to DNP to reappear at 4° C is probably explained by the requirement of metabolic processes for the turnover of surface Ig[9], and not by pinocytosis and modulation of receptors, which did not occur at 4° C[20].

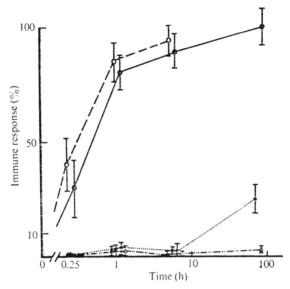

Fig. 2 *In vitro* AFC response of CBA spleen cells to pulsed exposures of immunogenic concentrations of $DNP_{1.5}$ Fla, $DNP_{12}HGG$ and $DNP_{12}FGG$. Results pooled from three experiments ±s.e. (8–12 cultures per point) abscissa-time of incubation with antigen. The total period of culture was 96 h in all cases. ●, DNP-Fla 37° C; ○, DNP-Fla 4° C; × ... ×, DNP-HGG 37° C; * ... *, DNP-HGG 4° C; △—··—△, DNP-FGG 37° C; ▲—··—▲, DNP-FGG 4° C.

Rosettes did not recover in cells pretreated with DNP Fla. At 4° C, the antigen must still be present on the cell surface. DNP-Fla which is very long[11,12] must be tethered to the cell surface by bonds at many different receptor antibody molecules, and unless these receptors are released synchronously at 37° C, the DNP-Fla could not leave the cell surface. The slight, yet significant ($P<0.0025$) degree of recovery of RFC would be consistent with this concept, but does not exclude the possibility that cap formation and pinocytosis of receptor-antigen complexes occurred as proposed by Taylor *et al.*[21], provided that this process was not complete. Obviously the simple protocol of these experiments cannot indicate which of these processes predominates. Whatever the receptor perturbations occurring on the B cell surface, the immunogenic correlates are clear. The firm binding of DNP-Fla (or DNP POL) correlates

65

very well with its marked *in vitro* immunogenicity, even with a transient exposure to cells (Fig. 2). It also explains why these polymeric antigens immunize B cells without the usual helper mechanisms[6,7] which nonpolymeric antigens require, since at 37° C $DNP_{12}HGG$ is rapidly lost from the B cell surface. Whatever is the mechanisms of cell collaboration, it may simply be envisaged as providing a more rigid framework for the carrier molecule, mimicking the structure of T cell independent antigens and enabling the hapten to remain in contact with B cell receptors for a sufficient period of time to immunize. Recent evidence that macrophages are essential for collaborative responses *in vitro* (unpublished work by M. F.), and that T cells need not be in contact with B cells[21], suggests that the site of immune induction may be on the surface of phagocytes, for which B cells have a marked affinity[22].

J. D. W. holds a senior fellowship of the New Zealand Medical Research Council. M. F. is supported by a National Health and Medical Research Council fellowship. This work was supported by grants from the National Health and Medical Research Council and the Australian Research Grants Committee and from the US Public Health Service. We thank Mrs J. Thompson and Misses G. Horlock and S. Owen for technical assistance.

[1] Miller, J. F. A. P., and Mitchell, G. F., *Transplant. Rev.*, **1**, 3 (1969).
[2] Davies, A. J. S., *Transplant. Rev.*, **1**, 43 (1969).
[3] Claman, H. N., and Chaperon, E. A., *Transplant. Rev.*, **1**, 92 (1969).
[4] Taylor, R. B., *Transplant. Rev.*, **1**, 114 (1969).
[5] Miller, J. F. A. P., in *Cell Interactions and Receptor Antibodies in Immune Responses* (edit. by Makela, O., Cross, A., and Kodunen, T. U.), 293 (Academic Press, London and New York, 1971).
[6] Feldmann, M., and Basten, A., *J. Exp. Med.*, **134**, 103 (1971)
[7] Feldmann, M., *Europ. J. Immunol.* (in the press).
[8] Greaves, M. F., *Transplant. Rev.*, **5**, 45 (1970).
[9] Cone, R. E., Marchalonis, J. J., and Rolley, R. T., *J. Exp. Med.* (in the press).
[10] Raff, M., *Nature*, **224**, 378 (1969).
[11] Feldmann, M., *J. Exp. Med.* (in the press).
[12] Feldmann, M., *Nature New Biology*, **231**, 18 (1971).
[13] Vitetta, E. S., Baur, S., and Uhr, J. W., *J. Exp. Med.*, **134**, 242 (1971).
[14] Ada, G., Nossal, G. J. V., Pye, J., and Abbot, A., *Austral. J. Exp. Biol. Med. Sci.*, **42**, 267 (1964).
[15] Feldmann, M., and Diener, E., *Immunology*, **21**, 387 (1971).
[16] Wilson, J. D., *Immunology*, **21**, 233 (1971).
[17] Dutton, P. W., and Bulman, H. N., *Immunology*, **7**, 54 (1964).
[18] Eisen, H. N., and Siskind, G. W., *Biochemistry*, **3**, 996 (1964).
[19] Hornick, L. L., and Karush, F., *Israel. J. Med. Sci.*, **5**, 163 (1969).
[20] Taylor, R. B., Duffus, W. P. H., Raff, M., and de Petris, S., *Nature*, **233**, 225 (1971).
[21] Feldmann, M., and Basten, A., *Nature New Biology*, **237**, 13 (1972).
[22] Schmidtke, J., and Unanue, E. R., *Nature New Biology*, **233**, 84 (1971).

Cell Cooperation in Humoral Immune Responses

A. C. ALLISON

A. J. S. DAVIES

Requirement of Thymus-dependent Lymphocytes for Potentiation by Adjuvants of Antibody Formation

THE mode of action of immunological adjuvants is of theoretical as well as practical importance. Adjuvants can not only increase immune responses; they can also bring about change from one type of immune response to another. For example, doses of bovine serum albumin (BSA) that in the absence of adjuvant induce tolerance, in the presence of adjuvants induce antibody formation[1,2]. To understand how adjuvants switch on antibody formation it may be necessary to determine which of the cells involved in the immune response are affected by adjuvants. Earlier experiments[3-5] led to the conclusion that the cells initially involved in the action of adjuvants are macrophages. This interpretation was based on observations that particulate adjuvants (such as bacteria) are ingested by macrophages but not by lymphocytes. Non-particulate or particulate adjuvants taken up by macrophages in culture and injected into syngeneic mice increased the antibody response of the recipients to two proteins (*Maia squinada* haemocyanin or bovine serum albumin (BSA)); in contrast, adjuvants taken up by lymphoid cells used to reconstitute immune responses in irradiated recipients had no demonstrable effect on antibody formation.

It is now clear that immune responses to many antigens require co-operation of thymus-dependent (T) and bone marrow derived (B) lymphocytes[6], the former possibly functioning as an antigen-concentrating system while the latter give rise to cells which can synthesize and release antibody[7,8]. If it is accepted that adjuvants act initially on macrophages, which are not themselves immunocompetent, two hypotheses can be considered. According to the first, adjuvants increase the probability of direct interaction of macrophages containing antigen and B lymphocytes capable of producing antibody, by-passing the T lymphocytes. According to the second, adjuvants increase the effectiveness of the interaction of macrophages and T lymphocytes, and this results in greater stimulation of the appropriate B lymphocytes to produce specific antibody. In an attempt to decide between these hypotheses we have compared the effects of adjuvants on antibody production against BSA in normal young adult CBA male mice and CBA mice deprived of T lymphocytes by neonatal thymectomy followed by two injections at 5 and 6 weeks of

Table 1 Serum Antibody against Bovine Serum Albumin in Normal and T Cell-Deprived Mice with and without Adjuvants

Treatment	Restoration	Adjuvant	Primary ABC	s.e.	Secondary ABC	s.e.
None	—	—	-0.42	0.14	0.57	0.15
None	—	LPS	0.34	0.08	1.61	0.17
None	—	Pertussis	0.24	0.06	1.73	0.26
None	—	ICFA	0.38	0.14	1.32	0.19
None	—	CFA	0.83	0.18	1.62	0.23
None*	—	'Retinol'	1.23	0.22	2.17	0.26
TX, ALS	—	—	-1.04	0.12	-0.96	0.12
TX, ALS*	—	LPS	-1.15	0.23	-1.12	0.18
TX, ALS	—	Pertussis	-1.32	0.14	-0.83	0.25
TX, ALS	—	ICFA	-1.37	0.11	-0.37	0.12
TX, ALS	—	CFA	-1.22	0.06	-0.93	0.15
TX, ALS	—	'Retinol'	-1.49	0.13	-0.82	0.21
TX, X-ray	Marrow	—	-1.36	0.19	-1.06	0.09
TX, X-ray	Marrow	LPS	-1.19	0.17	-1.27	0.13
TX, X-ray	Marrow	Pertussis	-1.08	0.08	-1.56	0.24
TX, X-ray	Marrow, thymus	—	-0.22	0.15	0.21	0.10
TX, X-ray	Marrow, thymus	LPS	0.13	0.09	1.26	0.08
TX, X-ray*	Marrow, thymus	Pertussis	0.27	0.06	1.34	0.27

0.2 ml. of rabbit anti-mouse lymphocyte serum (ALS) in some experiments, or thymectomy at 8 weeks followed by 850 r. whole-body irradiation and reconstitution with syngeneic bone marrow[9]. As an additional control, thymectomized (TX), irradiated and reconstituted mice receiving also a syngeneic thymus graft under the renal capsule[10] have been used.

The primary immunization was by intraperitoneal inoculation with 1 mg centrifuged BSA in 0.5 ml. saline; 20 days later the mice were bled for determination of primary antibody responses. The same day they were injected with 100 µg of BSA intraperitoneally, and blood samples were taken 10 days later for determination of secondary antibody responses. As adjuvants 2 µg of purified E. coli lipopolysaccharide (LPS) or 0.4 µg of Bordetella pertussis vaccine (Wellcome) were administered together with the antigen in saline. Freund's incomplete adjuvant (ICFA—mineral oil, Bayo F, containing mannoside—mono-oleate, Difco), or Freund's complete adjuvant (FCA–ICFA with 100 µg of Myobacterium butyricum/ml.) or 5 mg of 'Retinol' in ICFA were injected intraperitoneally 1 h after administration of antigen in saline. Antibody was assayed by the Farr test[1,4]. The antigen-binding capacity for each group of six mice (except for those marked with an asterisk, which contained 5) is presented as the geometric mean of the antigen-binding capacity (ABC) expressed as $\log_{10}$, together with the standard error (s.e.) calculated from the variance.

Representative results are shown in Table 1. In mice depleted of T cells by either procedure antibody responses to BSA are decreased, as previously described by Taylor[11], and the disparity between normal and T cell-deprived mice is greater when the primary immunization is made with adjuvant. In mice reconstituted with a thymus graft marked stimulation of antibody formation by adjuvants is again seen.

It therefore seems that T lymphocytes are required for potentiation of antibody formation against BSA by adjuvants. It has been suggested[3-5] that one of the effects of adjuvants may be to bring about the release from macrophages of a factor which stimulates proliferation of lymphocytes, and that when antigen is presented at this time there is an increased probability of antibody formation rather than tolerance induction. In keeping with this, it has been found[12] that vitamin A, which is an efficient adjuvant but is not in itself immunogenic, produces marked blast transformation and proliferation of cells in the thymus-dependent areas of draining lymph nodes, whereas substances, such as paraffin oil, without adjuvant activity have at most slight effects on the thymus-dependent areas of lymph nodes. These findings are consistent with the possibility that one of the effects of adjuvants is to increase the effectiveness of the interaction of macrophages and T lymphocytes. The macrophages concentrate antigen, which is presented to T lymphocytes in highly immunogenic form along with a stimulus for the latter to proliferate. This greatly increases the probability that T lymphocytes will mount an immune response, which in turn allows them to co-operate in antibody production by B lymphocytes. But whether or not the interaction of macrophages with T cells is augmented by adjuvants it seems that stimulation of T cells is an important aspect of the mechanism of action of at least some adjuvants. It is of interest to note that injections of allogeneic cells may also provide a general stimulus to T cell proliferation and overcome the requirement for co-operation by populations of T cells specifically stimulated by antigen in the potentiation of an immune response[13].

Part of this work was supported by grants to the Chester Beatty Research Institute from the Medical Research Council and the Cancer Research Campaign.

[1] Mitchison, N. A., *Proc. Roy. Soc.*, B, **161**, 275 (1964).
[2] Dresser, D. W., and Mitchison, N. A., *Adv. Immunol.*, **8**, 129 (1968).
[3] Unanue, E. R., Askonas, B. A., and Allison, A. C., *J. Immunol.*, **103**, 71 (1969).
[4] Spitznagel, J. K., and Allison, A. C., *J. Immunol.*, **104**, 119 (1970).
[5] Spitznagel, J. K., and Allison, A. C., *J. Immunol.*, **104**, 128 (1970).
[6] Roitt, I. M., Greaves, M. F., Rorrigiani, G., Brostoff, J., and Playfair, J. H. L., *Lancet*, ii, 367 (1969).
[7] Mitchell, G. F., and Miller, J. F. A. P., *J. Exp. Med.*, **128**, 821 (1968).
[8] Mitchison, N. A., in *Sixth Intern. Immunopathol. Symp.* (edit. by Miescher, P.) (Schwabe, Basel, 1970).
[9] Davies, A. J. S., Leuchars, E., Wallis, V., and Koller, P. C., *Transplantation*, **4**, 438 (1966).
[10] Leuchars, E., Cross, A. M., and Dukor, P., *Transplantation*, **3**, 28 (1965).
[11] Taylor, R. B., *Transplant. Rev.*, **1**, 114 (1969).
[12] Taub, R. N., Krantz, A. R., and Dresser, D. W., *Immunology*, **18**, 171 (1970).
[13] Benacerraf, B., in *Sixth Intern. Immunopathol. Symp.* (edit. by Miescher, P.) (Schwabe, Basel, 1970).

Thymic Antigen-Reactive Cells Do Not Specify Serological Properties of Antibody [1]

FRANK R. ORSINI AND GUSTAVO CUDKOWICZ

Irradiated mice of the $(C3H \times C57BL/10)F_1$ and $(C57BL/6 \times DBA/2)F_1$ strains were reconstituted with an excess of syngeneic bone marrow cells containing precursors of immunocytes, and with graded limiting numbers of thymocytes containing antigen-reactive cells (ARC), and then injected with sheep erythrocytes. The number of ARC and their possible specialization for serological properties of antibody were investigated by determining the titer of 2-mercaptoethanol-sensitive serum hemagglutinins and hemolysins 11 days after grafting. The limiting dilution assays indicated that the number of detectable $ARC/10^6$ thymocytes was of the same order of magnitude for both antibody responses. Agglutinins and lysins were associated in most recipient mice receiving an average of 1 ARC. Hence, serological properties of antibodies were not dictated by ARC, but by other cells participating in the immune responses, presumably of nonthymic origin.

Antibody-forming cells originate from bone marrow-derived precursors during primary and secondary immune responses (1, 2). Proliferation, differentiation, and maturation of potentially immunocompetent precursors into immunocytes are triggered by antigen and usually require the cooperation of accessory cells, e.g., thymus-derived inducers (1, 2) and marrow-derived macrophage-like cells (3, 4). Even though the inducer cells are antigen specific (5, 6), they do not dictate the specificity and class of antibody to be synthesized in immune responses to sheep erythrocytes (7–9) and to haptens (10). Marrow precursors of immunocytes are determined for the molecular class (11, 12) and specificity (8, 9) of antibody prior to interaction with thymus-derived cells and contact with administered antigen. The experiments reported in this paper show that thymic antigen-reactive cells also fail to dictate the serological properties of anti-sheep erythrocyte antibodies. Hemagglutinins and hemolysins are both produced under conditions in which thymus-derived, but not marrow-derived, cells limit the immune responses. Thus, the serological properties of emerging antibodies are determined by cells other than those of thymic origin.

[1] Research supported by U.S. Public Health Service Grant AM-13,969 from the Institute of Arthritis and Metabolic Diseases, and by American Cancer Society Grant T-476.

MATERIALS AND METHODS

Mice. (C3H/He×C57BL/10Cz)F$_1$ and (C57BL/6Cr×DBA/2Cr)F$_1$ female mice were used as donors and recipients in syngeneic cell transfers. The abbreviated designations are C3BF$_1$ and BDF$_1$, respectively. C3BF$_1$ mice were obtained from the West Seneca Animal Production Unit of Roswell Park Memorial Institute, Buffalo, N. Y., and BDF$_1$ mice from the Animal Production Section, National Cancer Institute, NIH, Bethesda, Md. The animals were used as recipients at the age of 10–12 weeks and as donors at the age of 8–10 weeks.

Irradiation. Mice to be grafted with bone marrow and thymus cells were exposed to 900 rads of ^{137}Cs gamma radiation from two opposite sources, at the rate of 127 rads/min, using a Gammacell-20 irradiator (Atomic Energy of Canada, Ltd., Ottawa, Canada). Radiation was given a few hours before transplantation.

Cell suspensions and transplantation. Nucleated bone marrow cells and thymocytes of normal mice were suspended separately in Eagle's medium, counted, mixed *in vitro,* and immediately injected into a lateral tail vein of irradiated mice, as previously described (13).

Immunization. After transplantation of nucleated cells, each mouse received either one or two intravenous injections of 5×10^8 washed sheep erythrocytes (SRBC) in 0.5-ml volumes; the mice immunized once were given SRBC on day 1, and those immunized twice on days 1 and 4. Fresh erythrocytes in Alsever's solution were purchased every second week from Gibco Microbio Lab, Madison, Wis. SRBC were washed three times in calcium and magnesium-free phosphate-buffered saline, pH 7.2, and suspended in Eagle's medium.

Assays for serum hemagglutinins and hemolysins. Blood was obtained from the retroorbital sinus and the sera were stored at $-20°$. Before titrations, the sera were inactivated at 56° for 30 min. Diluents for hemagglutination tests were 1% heat-inactivated guinea pig serum in phosphate-buffered saline (14); and for hemolysis tests, Hanks' balanced salt solution without phenol red. Titrations were done with 0.025-ml amounts of twofold serial dilutions of immune sera and an equal volume of 1% suspensions of SRBC using the Microtiter apparatus (Cooke Engineering Co., Alexandria, Va.). For the hemolysis test, 0.025 ml of complement (1:10 dilution of guinea pig serum, Gibco, Grand Island, N. Y.) was also added. The Microtiter trays (U-shaped wells for agglutination and V-shaped wells for hemolysis) were incubated for 1 hr at 37° and the patterns of hemagglutination, the hemolysis, and the endpoints ($\log_2$) were recorded immediately. The highest dilution of antiserum showing agglutination and the dilution showing ~50% lysis were used to indicate hemagglutination and hemolysis titers, respectively.

2-Mercaptoethanol (Eastman Organic Corporation, Rochester, N. Y.) was prepared as a 0.2 M solution in phosphate-buffered saline, pH 7.2, and added to an equal volume of undiluted serum. After 1 hr of incubation at 37° the sera were titrated for mercaptoethanol (ME)-resistant antibodies (15), using the techniques described above. To augment or facilitate ME-resistant antibodies, 0.025 ml goat anti-mouse gamma globulin serum of known activity (Hyland Division, Travenol Laboratories, Inc., Los Angeles, Calif.), diluted 1:20 in phosphate-buffered saline, was added to each well containing immune sera and SRBC (16). The reagents were mixed in the following order: test sera and SRBC, anti-mouse serum 10 min

73

later, and complement (for hemolysis only) 5 min later. Standard controls, such as substituting diluent for test sera, ME, and goat anti-mouse serum, were set up with each group of titrations.

Statistical methods. The Poisson model was used to predict the probability that antigen-reactive cells (ARC) contained in a given number of transplanted thymocytes reach the sites of antibody formation, interact with antigen, and induce marrow-derived cells to synthesize hemagglutinins and hemolysins. The method of maximum likelihood was used to estimate the probability values and their 95% confidence intervals (17).

Chi-square tests for independence were done on the two antibody responses of selected groups of mice to determine whether individual ARC could have induced the synthesis of both agglutinins and lysins.

Abbreviations used: ARC, antigen-reactive cells of thymic origin; ASU, antigen-sensitive unit; ME, 2-mercaptoethanol; P-AFC, precursors of antibody-forming cells of marrow origin; SIC, specific inducer cells of thymic origin; SRBC, sheep erythrocytes.

RESULTS

Experimental design. The experimental system used resembled the one described in previous publications (6, 7). Irradiated mice were transplanted with a fixed large number of bone marrow cells, which provided an excess of precursors of antibody-forming cells (P-AFC), together with graded and small numbers of thymocytes providing ARC. To induce antibody formation by the grafted cells, SRBC were given after irradiation and transplantation. The host animals contributed radioresistant phagocytic cells endowed with helper functions required for anti-SRBC responses (18). These cells are probably comparable to the macrophage-like radioresistant cells participating in immune responses *in vitro* (4, 19). The number of thymocytes mixed with bone marrow cells was reduced by twofold serial dilutions so as to limit the number of available ARC to one, a few, or none. Under these conditions, specialization of ARC for serological properties of antibody would result in independent assortment of hemagglutinin and hemolysin responses in groups of recipient mice. If, however, ARC were not specialized or restricted, both antibody responses would occur in individual mice once a single ARC is stimulated by antigen in the presence of an excess of marrow cells.

The following questions were asked: (a) can one determine by limiting dilution assays and serum titrations the number of thymocytes containing one detectable ARC? (b) Are these numbers the same or different for hemagglutinins and hemolysins? ((c) Assuming that the concentration of ARC is the same for the two responses, do individual ARC induce both responses or only one of them?

Immune responses in control mice. A group of 30 irradiated C3BF$_1$ mice was grafted with 5×10^7 bone marrow cells and 5×10^7 thymocytes and injected with SRBC on day 1. Five different mice were bled every second day from the day 5 to 35 after transplantation to determine serum antibody titers; the interval between bleedings was not shorter than 12 days for each mouse. By day 9 after transplantation, the sera of all recipient mice had a titer for ME-sensitive hemagglutinins and hemolysins, and by day 11 these titers ($\log_2$) reached peak values, ranging from 3

74

to 5. Low titers of ME-resistant antibodies appeared in a small fraction of mice bled on day 13, but the fraction of positive mice gradually increased and titers reached peak values on day 25. For ME-resistant hemagglutinins, all mice were positive on day 25, until the end of the experiment. The log_2 titers ranged from 3 to 6. In contrast, only 50% of the animals became positive for ME-resistant hemolysins and the log_2 titers were lower, from 2 to 5.

A total of 30 recipients injected with marrow cells and SRBC, but not with thymocytes, or with marrow-thymus cell mixtures, but not with SRBC, were set up as negative controls. In all but two mice the log_2 titers were <1 for hemagglutinins and hemolysins, and in the two exceptional mice the titers were 1.

Frequencies of anti-SRBC responses in C3BF₁ mice. In repeated experiments 114 irradiated mice were injected with 5×10^7 marrow cells and graded numbers of thymocytes ($1.5–50 \times 10^6$). SRBC were injected 1 day later. Blood samples were taken on day 11, the time of peak ME-sensitive antibody responses. The sera were titrated for hemagglutinins and hemolysins. All positive sera had log_2 titers ranging from 2 to 6, and the negative sera had titers of 1 or <1 (Table 1). As the number of thymocytes injected increased, the proportion of mice with positive sera also increased for both antibody responses. The relationship conformed to predictions of the Poisson model (Fig. 1). The numbers of thymocytes required for $\sim 63\%$ of the recipient sera to be positive (i.e., the inoculum sizes containing one detectable ARC) were similar for hemagglutinins and hemolysins (Table 1), although slightly more cells were necessary for hemolysin responses. The differences were not statistically significant, and for this reason subsequent calculations using the

TABLE 1

PERCENTAGE OF POSITIVE SERA IN IRRADIATED C3BF₁ MICE 11 DAYS AFTER INJECTION OF 5×10^7 MARROW CELLS, GRADED NUMBERS OF THYMOCYTES, AND SRBC

No. of thymocytes transplanted	Fraction of positive sera [a]	Percentage of positive sera [a]	Probability of positive sera per 10^6 cells [b]	Frequency of detectable ARC [b]
$\times 10^6$				$\times 10^{-6}$
	Hemagglutinins			
1.56	1/16	6.2		
3.12	4/20	20.0		
6.25	14/20	70.0	0.12	1/8.62
12.50	16/20	80.0	(0.08–0.16)	(1/6.45–1/11.76)
25.00	19/20	95.0		
50.00	18/18	100.0		
	Hemolysins			
1.56	0/16	0.0		
3.12	4/20	20.0		
6.25	13/20	65.0	0.09	1/11.24
12.50	12/20	60.0	(0.06–0.12)	(1/8.40–1/15.38)
25.00	18/20	90.0		
50.00	18/18	100.0		

[a] Log_2 titers 2–6.
[b] Calculated according to Ref. 17, with 95% confidence intervals in parentheses.

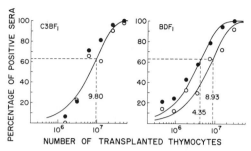

Fig. 1. Percentage of recipient mice with sera positive for anti-SRBC hemagglutinins (closed symbols) and hemolysins (open symbols) 11 days after transplantation of large numbers of bone marrow cells and limiting numbers of thymocytes. Symbols indicate observed percentages, and the fitted curves expected percentages according to the Poisson model. Dashed straight lines indicate the number of thymocytes containing an average of one detectable ARC. For C3BF$_1$ mice means of pooled percentages of positive sera for hemagglutinins and hemolysins were used.

Poisson model were made on pooled data (Fig. 1). The best estimate of the frequency of ARC detectable by the methods used was $1/9.8 \times 10^6$ thymocytes, with 95% confidence intervals of $7.3–13.3 \times 10^6$ thymocytes.

Most sera were either positive for hemagglutinins and hemolysins or negative for both (Table 3). Consequently, the results of chi-square tests were not compatible with the assumption that the two responses assorted independently. It rather appeared that individual ARC were capable of inducing both responses. The chi-square tests were made on data from groups of mice in which the proportion of positive or negative sera were not extremely high or low.

Frequencies of anti-SRBC responses in BDF$_1$ mice. In several repeated experiments 401 irradiated mice were injected with either 2 or 6×10^7 marrow cells and graded numbers of thymocytes ($0.47–30 \times 10^6$). SRBC were injected 1 and 4 days later. Eleven-day titers of hemagglutinins and hemolysins were determined, and the sera were classified as positive or negative. Values of $\log_2$ titers, as well as relations between the $\log_{10}$ numbers of thymocytes grafted and the percentages of responses, did not differ greatly in this strain from those in C3BF$_1$ mice (Table 2 and Fig. 1). However, hemolysin responses were less frequent than hemagglutinin responses, particularly in mice injected with 2×10^7 marrow cells. The titers reflected ME-sensitive antibodies since ME-resistant titers were very low and rare (3 of 131 mice) and goat anti-mouse sera failed to augment. The frequency of ARC measured by hemagglutinin responses was $1/4.35 \times 10^6$ thymocytes, whereas that measured by hemolysin responses was $1/8.93$. The difference of these values was statistically significant (Table 2) but due primarily to the fewer positive hemolytic sera of mice given 2×10^7 marrow cells. In additional experiments with 6×10^7 marrow cells the numbers of hemagglutinin and hemolysin responses were, in fact, nearly equal. This suggested that bone marrow cells *and* thymocytes limited hemolysin responses in most of the experiments with BDF$_1$ mice, and that the hemolysis test might be less sensitive than the hemagglutination test. If the differences in values of ARC frequencies were apparent rather than real, individual ARC of BDF$_1$ thymuses should have induced both antibodies as did ARC of C3BF$_1$ thymuses.

76

TABLE 2

PERCENTAGE OF POSITIVE SERA IN IRRADIATED BDF_1 MICE 11 DAYS AFTER
INJECTION OF 2 OR 6×10^7 MARROW CELLS, GRADED
NUMBERS OF THYMOCYTES, AND SRBC

No. of thymocytes transplanted	Fraction of positive sera [a]	Percentage of positive sera [a]	Probability of positive sera per 10^6 cells [b]	Frequency of detectable ARC [b]
$\times 10^6$				$\times 10^{-6}$
	Hemagglutinins			
0.47	3/14	21.4		
0.94	15/67 [c]	24.4		
	5/15			
1.88	28/64	42.8		
	5/13			
3.75	33/63	57.1	0.23	1/4.35
	11/14		(0.20–0.27)	(1/3.72–1/5.10)
7.50	35/45	78.3		
	12/15			
15.00	38/40	94.4		
	13/14			
30.00	25/25	100.0		
	12/12			
	Hemolysins			
0.47	2/14	14.3		
0.94	7/67 [c]	12.2		
	3/15			
1.88	20/64	32.5		
	5/13			
3.75	17/63	29.9	0.11	1/8.93
	6/14		(0.09–0.13)	(1/7.58–1/10.64)
7.50	27/45	63.3		
	11/15			
15.00	27/40	72.2		
	12/14			
30.00	23/25	91.9		
	11/12			

[a] Log_2 titers 2–6.
[b] Calculated according to Ref. 17, with 95% confidence intervals in parentheses.
[c] The first line indicates mice injected with 2×10^7 marrow cells and the second line those injected with 6×10^7 cells. Percentage values were calculated on all mice.

Hemagglutinin and hemolysin responses were associated in the majority of mice injected with limiting numbers of thymocytes. Results of chi-square tests were not compatible with the assumption that separate ARC induced hemagglutinins and hemolysins independently (Table 3). The more reliable estimate of ARC frequency appears, therefore, to be the one derived from hemagglutinin responses.

DISCUSSION

The experiments described answered the three questions asked at the onset of this article. The frequencies of ARC in thymocyte suspensions of two donor strains were determined by limiting dilutions for anti-SRBC humoral antibody responses.

TABLE 3

CHI-SQUARE TEST FOR INDEPENDENCE OF HEMAGGLUTININ AND HEMOLYSIN
RESPONSES IN RECIPIENTS OF LIMITING NUMBERS OF THYMOCYTES

Hemag-glutinins	Hemol-ysins	No. of thymocytes ($\times 10^6$)							
		C3BF$_1$			BDF$_1$				
		3.12 (20)[a]	6.25 (20)	12.50 (20)	0.94 (82)	1.88 (77)	3.75 (77)	7.50 (60)	15.00 (54)
+[b]	+	3	13	12	6	24	22	35	39
+	−[c]	1	1	4	14	10	22	14	12
−	+	1	0	0	5	3	1	3	0
−	−	15	6	4	57	40	32	8	3
χ^2 [d]		5.64	12.10	4.70	4.52	31.01	17.68	5.76	4.89

[a] Numbers in parentheses indicate number of mice
[b] +, positive sera.
[c] −, negative sera.
[d] Chi-square values in the table were compared with 3.84 and 6.63, the χ^2 statistics at 0.05 and 0.01 levels of significance, respectively. All values were incompatible with the hypothesis that hemagglutinin and hemolysin responses were independent of each other.

The two strains were known to differ with respect to the frequency of thymic ARC from previous limiting dilution studies for anti-SRBC hemolytic plaque-forming cells in the spleen (6, 7). The estimates were in close agreement regardless of whether serum hemagglutinins, hemolysins (in C3BF$_1$ mice), or splenic plaque-forming cells of the direct and indirect classes (6, 7) were assessed. The 95% confidence intervals of the estimated frequencies of ARC overlapped for any of the four techniques used, except for hemolysin responses in BDF$_1$ mice not given sufficient bone marrow. ARC were ~3 times more numerous in BDF$_1$ than in C3BF$_1$ thymocytes. The fact that not more ARC were detected by serum antibodies than by splenic plaque-forming cells indicated that most or all ARC retaining function upon transplantation were those which settled in the recipient spleens. The independent confirmation of previous estimates of ARC frequency added confidence in the limiting dilution approach. The procedure was not only accurate and reproducible in enumerating cells limiting immune responses, but also useful for elucidating functions. The interpretation, limitations, and necessary controls of limiting dilution assays were discussed elsewhere (20, 21). It is to be noted, however, that estimates of ARC were obtained so far from tests done 9–11 days after cell transplantation and immunization, but not later.

In interpreting the data presented in this paper, one must consider that SRBC are a complex mosaic of antigenic determinants and that hemagglutinins and hemolysins could represent antibodies with different specificities. On the other hand, it has been shown that subclasses of guinea pig (22) and mouse (23, 24) IgG antibodies possess different biological and serological properties. In particular, the hemagglutinating and complement-dependent hemolytic activities were not always associated in subclasses of antihapten antibodies. The present experiments were concerned with the possible determination by thymic ARC of the ability of ME-sensitive antibody to agglutinate or to lyse erythrocytes *in vitro*. A structural difference of the antibodies with the two serological properties could depend on the presence of a complement-fixing site on the Fc fragment of hemolysins, and the absence of this site from hemagglutinins. Thus, serological properties of antibodies may reflect a kind of heterogeneity not due to differences in molecular class or subclass, specificity, and idiotypes or allotypes.

ME-sensitive hemagglutinins and hemolysins were associated in sera of mice grafted with limiting numbers of thymocytes. Since not more than one or two ARC should have been functional in most mice with positive sera, association of the antibody responses was interpreted to mean that individual ARC were not restricted to participate in interactions leading exclusively to antibody of one or the other serological type. ARC proliferate extensively in response to antigen administration and generate specific inducer cells (SIC) (6, 25). Therefore, clonal populations of SIC must have interacted with P-AFC and/or other cells responsible for the synthesis of hemagglutinins and hemolysins. In view of the restriction of splenic antigen-sensitive units (ASU) for hemagglutinins and hemolysins (13, 20, 21), marrow-derived cells could possibly have determined the serological properties of antibody. The data presented exclude thymic ARC as the source of information for synthesis of ME-sensitive agglutinins versus lysins, although their inductive function was necessary for the synthesis of both antibodies.

Immune responses to antigens of SRBC require at least three distinct cell types (1–5) which mature and integrate into ASU in the peripheral lymphoid organs of mice (13, 20, 21). The roles of cells constituting ASU are known to a limited extent: bone marrow-derived cells produce antibody (1, 2, 5) and possess the information for molecular class and specificity (8–12). Thymus-derived cells do not synthesize antibodies (25, 26) but they may facilitate recognition of antigenic determinants since they possess information for specificity (5, 6). The facts that ARC lack specialization for molecular class (7) and serological properties of antibody, and the antigen specificity of P-AFC, point to a role other than transfer of information by thymus-derived cells to P-AFC.

REFERENCES

1. Mitchell, G. F., and Miller, J. F. A. P., *J. Exp. Med* **128**, 821, 1968.
2. Jacobson, E. B., L'Age-Stehr, J., and Herzenberg, L. A., *J. Exp. Med.* **131**, 1109, 1970.
3. Mosier, D. E., and Coppleson, L. W., *Proc. Nat. Acad. Sci. U.S.A.* **61**, 542, 1968.
4. Haskill, J. S., Byrt, P., and Marbrook, J., *J. Exp. Med.* **131**, 57, 1970.
5. Mitchell, G. F., and Miller, J. F. A. P., *Proc. Nat. Acad. Sci. U.S.A.* **59**, 296, 1968.
6. Shearer, G. M., and Cudkowicz, G., *J. Exp. Med.* **130**, 1243, 1969.
7. Shearer, G. M., Cudkowicz, G., and Priore, R. L., *J. Exp. Med.* **130**, 467, 1969.
8. Miller, H. C., and Cudkowicz, G., *J. Exp. Med* **132**, 1122, 1970.
9. Miller, H. C., and Cudkowicz, G., *J. Exp. Med.* **133**, 1971.
10. Schirrmacher, V., and Rajewsky, K., *J. Exp. Med.* **132**, 1019, 1970.
11. Cudkowicz, G., Shearer, G. M., and Ito, T., *J. Exp. Med.* **132**, 623, 1970.
12. Miller, H. C., and Cudkowicz, G., *Science* **171**, 913, 1971.
13. Shearer, G. M., and Cudkowicz, G., *J. Exp. Med.* **129**, 935, 1969.
14. Golub, E. S., Mishell, R. I., Weigle, W. O., and Dutton, R. W., *J. Immunol.* **100**, 133, 1968.
15. Twarog, F. J., and Rose, N. R., *J. Immunol.* **102**, 375, 1969.
16. Rabin, B. S., and Rose, N. R., *Immunology,* **19**, 239, 1970.

79

17. Porter, E. H., and Berry, R. J., *Brit. J. Cancer* **17**, 583, 1963.
18. Miller, H. C., and Cudkowicz, G., Manuscript in preparation.
19. Roseman, J. M., *Science* **165**, 1125, 1969.
20. Shearer, G. M., Cudkowicz, G., Connell, M. St. J., and Priore, R. L., *J. Exp. Med.* **128**, 437, 1968.
21. Shearer, G. M., Cudkowicz, G., and Priore, R. L., *J. Exp. Med.* **129**, 185, 1969.
22. Bloch, K. J., Kourilsky, F. M., Ovary, Z., and Benacerraf, B., *J. Exp. Med.* **117**, 965, 1963.
23. Nussenzweig, R. S., Merryman, C., and Benacerraf, B., *J. Exp. Med.* **120**, 315, 1964.
24. Voisin, G. A., Kinsky, R. G., and Jansen, F. K., *Nature London* **210**, 138, 1966.
25. Davies, A. J. S., Leuchars, E., Wallis, V., and Koller, P. C., *Transplantation* **4**, 438, 1966.
26. Davies, A. J. S., Leuchars, E., Wallis, V., Marchant, R., and Elliott, E. V., *Transplantation* **5**, 222, 1967.

IN VITRO COOPERATION OF CELLS OF BONE MARROW AND THYMUS ORIGINS IN THE GENERATION OF ANTIBODY-FORMING CELLS[1]

DOUGLAS C. VANN[2] AND JOHN R. KETTMAN[3]

B cells and T cells cooperate in the *in vitro* generation of hemolytic plaque-forming cells to sheep, horse and burro erythrocytes. The B cells are obtained from the spleens of lethally irradiated mice injected with bone marrow cells. The T cells are obtained from the spleens of lethally irradiated mice injected with thymus cells and heterologous erythrocytes. Both cell populations are obtained after 7 days residence in the irradiated hosts. Neither cell population can respond when cultured alone. The *in vitro* function of B cells, but not T cells, is inactivated by exposure to 2000 R γ irradiation. The T cells function best in *in vitro* response to the erythrocyte antigens with which they were injected into the irradiated hosts. Responses were also generated to other foreign erythrocytes and to the trinitrophenyl hapten coupled to erythrocyte carriers.

Considerable evidence for cellular cooperative interactions in the antibody responses of mice has accumulated in the past few years. These results enable the construction of a generalized picture of the cellular basis of antibody production to heterologous erythrocyte (RBC) antigens. It is currently thought that precursors to antibody-producing cells can be found in the bone marrow (B cells) and, in the presence of antigen-specific cells derived from the thymus (T cells), appropriate antigens will stimulate the precursors to proliferate and differentiate into antibody-producing cells. The mechanism of this cooperation is unclear, but the cellular roles are well substantiated by evidence from a number of experiments. Claman *et al.* (1) and Mitchell and

Miller (2) have demonstrated that both B cells and T cells are required to regenerate immunocompetence in irradiated syngeneic recipients. Davies *et al.* (3) found that although antigen induced cellular proliferation in thymus-derived cells, no antibody production ensued. The origin of hemolytic plaque-forming cells (PFC) from B cells was demonstrated on the basis of histocompatibility antigens by Mitchell and Miller (4) and on the basis of marker chromosomes by Nossal *et al.* (5). The antigen specificity of T cell function is inferred from double transfer experiments (Mitchell and Miller (2), Shearer and Cudkowicz (6), and Claman and Chaperon (7)). These experiments demonstrated that in order to recover functional T cells from the spleens of irradiated mice, thymus cells and antigen had to be injected into the animals. This effect was found to be antigen-specific in that the antigen administered with the thymus cells had to be the same as the challenging antigen for antibody production. These cells will be henceforth referred to as "educated" T cells.

Studies on antibody responses generated *in vitro* have implicated the need for a third kind of cell, one which behaves like a macrophage. This requirement is derived from studies on the physical fractionation of immunocompetent cell populations on the basis of the ability to adhere to surfaces (8, 9) or density fractionation by

[1] This is Publication No. 539 from the Department of Experimental Pathology, Scripps Clinic and Research Foundation, La Jolla, California 92037. This work was supported by grants from United States Public Health Service, Training Grant GM00683, AI-08795-02, and American Cancer Society, E-395D and Cancer Research Coordinating Committee, University of California.

[2] Supported by United States Public Health Service Training Grant GM00683. Present address: Department of Genetics, School of Medicine, University of Hawaii, Honolulu, Hawaii 96822.

[3] The Department of Biology, University of California, San Diego, La Jolla, California 92037.

BSA-gradient centrifugation (10). The exact role of the macrophage-like cell remains undefined and the possibility remains that its requirement could be a nonspecific effect of *in vitro* culture (11, 12).

Still more evidence for cellular cooperation in immune responses comes from dose-response analysis in cell transfer systems (13), in culture *in vitro* (14), and in diffusion chamber cultures (15). Cooperative phenomena are also seen in the "carrier effect," where prior immunization of an animal with carrier antigen will enhance an anti-hapten response *in vivo* only if subsequent challenge is performed with the hapten coupled to the original priming carrier (16, 17). The parallel between T-cell function and the enhancing effect of carrier-primed lymphoid cells is strengthened by the report that carrier-primed cells are sensitive to cytotoxic treatment with antiserum to the thymus antigen θ in the presence of guinea pig complement (18, 19), and by the observations of Mitchison *et al.* (20), who obtained evidence that the cells responsible for carrier enhancement are of thymus origin.

We have been prompted to seek out a method of obtaining *in vitro* cooperation of B cells and T cells in the hope that isolated cultures may be more amenable than intact animals to the study of the mechanism of cellular cooperation in immune responses. Several experimental procedures are available for such studies. Cooperation has been reported between spleen cells from neonatally thymectomized mice and irradiated normal spleen cells (21, 22) or thymus cells (23). Normal spleen cells show synergism with irradiated carrier-primed spleen cells in generating anti-hapten responses *in vitro* (24–26). A third approach has been to use educated T cells and B cells obtained from irradiated syngeneic recipients injected with bone marrow cells or from thymectomized, irradiated recipients of bone marrow cells. A preliminary report of this system has been presented (19) and Hartmann (27, 28) has also reported successful *in vitro* cooperation between educated T cells and B cells. Our results are in close accord with those of Hartmann and, as will be shown, we observed an apparent breakdown of antigen specificity for the T cell cooperative function *in vitro*. The question raised about the specificity of T cell cooperation prompted us to investigate the function of the T cells educated to carriers in generating responses to the hapten,

2,4,6-trinitrophenyl (TNP). We have also tested the sensitivity to radiation of B cell and educated T cell populations. This has provided information identifying the origin of antibody-producing cells generated *in vitro* and relating the activity of educated T cells to irradiated normal or carrier-primed spleen cells.

METHODS AND MATERIALS

Mice. BDF_1 (C57BL/6 female $\times$ DBA/2 male) hybrid mice which were bred in our own colony were used. In each experiment all mice were of the same sex.

Antigens. Erythrocytes from sheep (SRBC), horse (HRBC), or burro (BRBC) were washed in basal salts solution (BSS) for use as stimulatory antigens. TNP was coupled to RBC for assay by the method of Rittenberg and Pratt (29).

Irradiation. Mice were exposed to either 900 R of x-rays or 1050 to 1100 R from a ^{60}Co source. Dissociated cells received 2000 R from a ^{60}Co source while on ice, either in a cell pellet or suspended in BSS. Spleen cell responses were more efficiently suppressed when suspensions were irradiated.

Preparation of cells. Thymuses were removed from 8-week-old exsanguinated donors, stripped of connective tissue, and minced in BSS with scissors. Cells were freed from stromal tissue by repeated aspiration in a 2.5-ml plastic syringe. The cell suspensions were allowed to settle in 12-ml conical tubes for 3 to 5 min on ice and the supernatant single cell suspension decanted and centrifuged for 10 min at approximately 400 $\times$ G. The cells were resuspended in fresh BSS, counted, and adjusted to a concentration of 6 $\times$ 10^7 cells/ml. Heterologous RBC antigens were washed three times in BSS and a 5% v/v suspension prepared. Equal volumes of the thymus cell suspension and RBC suspensions were mixed and 1.0 ml of the mixture injected into the lateral tail vein of irradiated syngeneic recipient mice 2 to 3 hr after irradiation. The recipient mice were sacrificed 7 days later and single cell suspensions prepared from their spleens. The cells were washed and adjusted to a concentration of 10 $\times$ 10^6 cells/ml for *in vitro* culture. Thymus cells educated to SRBC, HRBC, or BRBC are denoted T_{SRBC}, T_{HRBC}, or T_{BRBC}, respectively.

Femurs and sometimes tibiae of 8-week-old donors were flushed with BSS to extrude bone marrow. Single cell suspensions were prepared by aspiration through a 20-gauge hypodermic needle. After settling to remove particles and debris, the cells were washed, counted and adjusted to 4×10^7 cells/ml. One milliliter of the cell suspension was injected i.v. into each irradiated recipient. After 7 days the recipients were killed, their spleens removed, and single cell suspensions prepared. The cells were washed, counted, and the concentration was adjusted to 10×10^6 cells/ml for in vitro culture. These cells also are referred to as B cells.

Culture and assay. The in vitro culture technique and modified Jerne hemolytic plaque assay have been described by Mishell and Dutton (30). Usually, each culture contained 10×10^6 cells in 1 ml of modified Eagle's medium containing 5% fetal bovine serum. Each line was run in triplicate cultures which were pooled at the time of assay. Anti-TNP responses were measured as described by Kettman and Dutton (31). All experiments reported were assayed on the 4th day of culture and the results expressed as PFC per culture. When the kinetics of the immune responses were determined, no different conclu-

sions were reached so only responses on the 4th day of culture are reported for clarity.

RESULTS

In vitro cooperation of B cells and educated T cells. B cells were cultured in vitro with T cells prepared in several ways and stimulated with SRBC and HRBC, as shown in Table I. It can be seen that all of the cell populations failed to generate significant PFC responses when cultured alone at a concentration of 10×10^6 cells per culture. When 5×10^6 B cells were cultured with 5×10^6 educated T cells, sizeable responses to SRBC and HRBC were detected. Cooperation between the two populations has usually been observed to generate at least 10 times as many PFC per culture as the sum of the responses of the two cell populations when cultured separately. In order to develop T cell activity, thymus cells and antigen had to be injected into the same irradiated recipient, as shown in lines I through M. If thymus cells were injected and antigen omitted or if thymus cells were omitted and antigen injected, the recipient spleens failed to cooperate with B cells in generating in vitro PFC

TABLE I

In vitro cooperation of B cells and "educated" T cells

Line	Cells Cultured[a]	Ag in Culture[b]	PFC per Culture Assayed with	
			SRBC	HRBC
A	B cells	S + H	37	0
B	T_{SRBC}	S + H	37	0
C	T_{HRBC}	S + H	0	0
D	$T_{nothing}$[c]	S + H	4	0
E	Nothing-SRBC[d]	S + H	0	0
F	Nothing-HRBC	S + H	4	0
G	B cells + T_{SRBC}	S + H	1340	248
H	B cells + T_{HRBC}	S + H	248	515
I	B cells + $T_{nothing}$	S + H	10	4
J	B cells + nothing-SRBC	S + H	37	10
K	B cells + nothing-HRBC	S + H	4	7
L	B cells + $T_{nothing}$ + nothing-SRBC	S + H	23	0
M	B cells + $T_{nothing}$ + nothing-HRBC	S + H	7	4

[a] 10×10^6 cells per dish when each kind of cell was cultured singly; 5×10^6 of each kind of cell cultured per dish for mixtures.

[b] Ag in culture: SRBC and HRBC stimulation of in vitro culture.

[c] $T_{nothing}$: thymus cells but no heterologous RBC injected into irradiated primary recipient.

[d] Nothing-SRBC and nothing-HRBC: heterologous RBC, but no thymus cells injected into irradiated primary recipient.

83

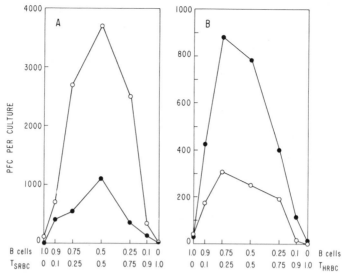

Figure 1. Effect of varying proportions of B cells and educated T cells. The cell concentration was kept constant at 10×10^6 cells per culture. *A*, cells cultured with T_{SRBC}; *B*, cells cultured with T_{HRBC}. All cultures stimulated *in vitro* with SRBC and HRBC. O——O, PFC assayed with SRBC; ●——●, PFC assayed with HRBC.

responses. The additional control of mixing two kinds of defective populations in culture with B cells also failed to facilitate production of PFC.

Optimal proportions of B cells and T cells. The proportions of B cells and educated T cells were varied, keeping the number of cells constant at 10×10^6 per culture. B cells were cultured with T_{SRBC} or T_{HRBC}, stimulated with SRBC and HRBC, and assayed on day 4 for PFC to SRBC or HRBC. The results are presented in Figure 1 and it can be seen that equal mixtures of B cells and T cells are approximately optimal for the generation of PFC responses.

Specificity of in vitro responses. A number of experiments testing the specificity of *in vitro* cooperation between B cells and educated T cells were performed. The results of two such experiments are shown in Table II. B cells were cultured with T cells obtained from irradiated recipients injected with thymus cells and SRBC, BRBC, HRBC. The antigen injected with the thymus cells will be referred to as the homologous antigen, and antigens not injected with the thymus cells termed heterologous antigens. No significant PFC responses occurred to antigens which were not present in the *in vitro* culture. Responses were

obtained with heterologous antigens, but showed variable patterns of responses. When SRBC was the heterologous antigen, significant responses to SRBC were only generated when the homologous antigen was also present in the culture (line F *vs* line G in Experiment A, and line G *vs* line H in Experiment B). When BRBC or HRBC were the heterologous antigens, responses to the heterologous antigen occurred in the absence of the homologous antigen, although a slight increase in the responses to HRBC was sometimes seen (line D *vs* line E in Experiment A, and line D *vs* line F in Experiment B). Cross-reactions between SRBC and BRBC (31) and between SRBC and HRBC have been found to be less than 1%. We have tested the PFC generated under the above conditions and found that less than 1% of the PFC produced clear plaques when mixtures of SRBC and HRBC were used as the target cells for the hemolytic plaque assay. The vast majority of cells produced hemolysins directed against non-cross-reacting determinants.

Anti-TNP responses. B cells were cultured *in vitro* with T cells educated to SRBC or HRBC and the cultures stimulated with heavily substituted TNP-SRBC and TNP-HRBC. PFC responses to native SRBC, HRBC, TNP-SRBC and TNP-HRBC were detected on day 4 as shown in Table III. These results demonstrate that *in vitro* cooperation between B cells and educated T cells can result in responses to determinants other than those to which the thymus cells were exposed while in residence in the irradiated animal.

Irradiation of B cells and T cells. Since cellular proliferation plays a large role in the generation of PFC (reviewed by Makinodan *et al.* (32)), it was reasoned that the population containing the

TABLE II

In vitro cooperation of B cells and "educated" T cells: specificity of responses

Line	Cells Cultured[a]	Ag in Culture[b]	PFC per Culture Assayed with		
			SRBC	BRBC	HRBC
		Experiment A			
A	B cells	S + B	60	0	
B	T$_{SRBC}$	S + B	0	0	
C	T$_{BRBC}$	S + B	0	0	
D	B cells + T$_{SRBC}$	S + B	4630	410	
E	B cells + T$_{SRBC}$	B	95	515	
F	B cells + T$_{BRBC}$	S + B	314	356	
G	B cells + T$_{BRBC}$	S	15	0	
		Experiment B			
A	B cells	S + H	20		7
B	T$_{SRBC}$	S + H	0		3
C	T$_{HRBC}$	S + H	3		0
D	B cells + T$_{SRBC}$	S + H	3618		409
E	B cells + T$_{SRBC}$	S	2680		57
F	B cells + T$_{SRBC}$	H	67		208
G	B cells + T$_{HRBC}$	S + H	429		1293
H	B cells + T$_{HRBC}$	S	94		0
I	B cells + T$_{HRBC}$	H	30		945

[a] 10×10^6 cells per dish when each kind of cells was cultured singly; 5×10^6 of each kind of cells cultured per dish for mixtures.

[b] Ag in culture: S, B and H indicate SRBC, BRBC and HRBC stimulation of *in vitro* cultures, respectively.

TABLE III

In vitro cooperation of B cells and "educated" T cells in the response to TNP[a]

Line	Cells Cultured[b, c]	PFC per Culture Assayed with					
		SRBC	TNP-SRBC	Δ[d]	HRBC	TNP-HRBC	Δ
A	B cells	170	312	142	20	80	60
B	T$_{SRBC}$	0	0	0	0	0	0
C	T$_{HRBC}$	0	0	0	0	0	0
D	B cells + T$_{SRBC}$	1310	2560	1250	300	680	380
E	B cells + T$_{HRBC}$	2180	9080	6900	1545	6840	5395

[a] A single experiment is reported, others confirm these results.

[b] 10×10^6 cells per dish when each kind of cells was cultured singly; 5×10^6 of each kind of cells cultured per dish for mixtures.

[c] All cultures with both antigens; highly substituted TNP-SRBC and TNP-HRBC.

[d] Δ, Difference between number of PFC when assayed with TNP coupled RBC and the number of PFC observed with native RBC.

TABLE IV
Effect of irradiation on B cell and "educated" T cell in vitro cooperation[a]

| Line | Cells Cultured[b] | Ag in Culture[c] | PFC per Culture Assayed with | |
			SRBC	HRBC
A	B cells	S + H	47	13
B	Irr. B cells[d]	S + H	10	0
C	T$_{SRBC}$	S + H	0	0
D	Irr. T$_{SRBC}$	S + H	0	0
E	T$_{HRBC}$	S + H	0	0
F	Irr. T$_{HRBC}$	S + H	0	0
G	B cells + T$_{SRBC}$	S	1120	
H	B cells + Irr. T$_{SRBC}$	S	1072	
I	B cells + T$_{HRBC}$	H		184
J	B cells + Irr. T$_{HRBC}$	H		144
K	Irr. B cells + T$_{SRBC}$	S + H	84	44
L	Irr. B cells + Irr. T$_{SRBC}$	S + H	44	10
M	Irr. B cells + T$_{HRBC}$	S + H	54	3
N	Irr. B cells + Irr. T$_{HRBC}$	S + H	7	0

[a] A single experiment is reported, others confirm these results.
[b] 10×10^6 cells per dish when each kind of cells was cultured singly; 5×10^6 of each kind of cells cultured per dish for mixtures.
[c] Ag in culture: SRBC and HRBC stimulation of *in vitro* cultures.
[d] Irradiated cells received 2000 r from a ^{60}Co source immediately before cultures were set up.

precursors to PFC arising from *in vitro* B cell and educated T cell cooperation should be radiosensitive. Aliquots of freshly prepared B cells, T$_{SRBC}$ and T$_{HRBC}$ were exposed to 2000 R from a ^{60}Co source at 0°C. Unirradiated aliquots were stored on ice for the same period of time. Equal numbers of B cells and educated T cells were cultured *in vitro* with antigenic stimulation at a total concentration of 10×10^6 cells per culture. As shown in Table IV, irradiation of the educated T cell populations had no effect on the ability to cooperate with unirradiated B cells in the generation of PFC responses. B cells were found to be radiosensitive and exposure to 2000 R reduced their ability to respond to background levels. This experiment demonstrates the radiosensitivity of B cells and the radioresistance of the function of educated T cell populations.

DISCUSSION

The results presented here demonstrate the cooperation of cells of bone-marrow and thymus origins in the Mishell-Dutton *in vitro* culture system. This experimental approach is useful for the study of B cell and T cell cooperation but it would not be expected to provide evidence about the nature of the third cell type, the macrophage-like cell capable of adhering to surfaces, which

has been implicated in the generation of antibody responses *in vitro* (8, 11). Because the adherent cell is radioresistant (33) it is expected that it could be found as a host contribution in either of the cooperating cell populations and has, in fact, been demonstrated in the spleens of irradiated recipients injected with bone marrow cells (34).

Preliminary experiments indicated that freshly prepared bone marrow and thymus cells failed to generate antibody responses in Mishell-Dutton cultures (P. A. Campbell, personal communication) or Millipore filter diffusion chamber cultures (D. C. Vann, unpublished results).

Our selection of the method of producing B cells by injecting of bone marrow cells into irradiated recipients and culturing the spleens of these animals 1 week later was based on the procedure described by Talmage *et al.* (34). We did not encounter the low responses reported by Hartmann (27, 28) when he used B cells obtained after 1 week's residence in irradiated animals.

The use of educated T cells was based on the experimental design of Mitchell and Miller (2) for the *in vivo* study of T cell and B cell cooperation which was further investigated by Shearer and Cudkowicz (6) and Claman and Chaperon (7). We hoped that since Shearer and Cudkowicz (6) had shown that the education process results in an increase in T cell activity, as measured by limiting dilution analysis, these cells would be able to function *in vitro*. Although there have been reports of cooperative activity of freshly prepared thymus cells *in vitro* (21, 23), Hartmann (28) found native thymus cells to have no effect or even to be deleterious to normal spleen responses *in vitro*. It is not known at this time whether the usefulness of educated T cells arises from an increase in the number of specific T cells, a change in the physiologic state of the T cells, or alterations in the composition of the T cell population. Our results confirm earlier observations (2, 6, 7) that thymus cells and antigen need to be injected into the same animal for the production of functional educated T cells. This has been interpreted to indicate that T cells are capable of recognizing different antigens, presumably by means of specific immunoglobulin receptors (35). Chan *et al.* (36) have demonstrated the sensitivity of educated T cells to exposure to antiserum to the thymus-specific antigen θ, in the presence of a source of complement. This cytotoxic treatment substantiates the idea that thymus-derived cells constitute the functional component of educated T cell populations.

Our observation of the breakdown of specificity of the response, as demonstrated by increased responses to heterologous antigens in the presence of the homologous antigen, appears to be similar to the phenomenon described by Hartmann (28). The significance of this effect is not clear, but it can be concluded that it is not due to cross-reactions at the level of the hemolysin-producing cells. It could be due to cross-reactions to shared antigenic determinants detected only by the T cells, but not the PFC. A second possibility is that T cells, upon reaction with their specific antigenic determinants, release a nonspecific stimulatory factor which enhances the generation of antibody-producing cells by B cells which have reacted with determinants recognized by the receptors on their surfaces. To explain why mixtures of B cells and T_{SRBC} give responses to heterologous antigens in the absence of the homologous antigen (SRBC), one could propose the presence of SRBC-like antigens in the fetal calf serum used in the culture system (31). In cell transfer experiments to irradiated recipients a similar nonspecific effect was not seen (D. C. Vann, unpublished observation).

Because of the question of specificity of B cell and educated T cell cooperation *in vitro*, experiments with the hapten, TNP, were performed to determine if the educated T cell population directed the response to antigenic determinants made by B cells. Our results indicate that responses can occur to determinants to which neither the T cell nor B cell populations have been previously exposed and indicate that educated T cells are analogous to carrier-primed spleen cell populations (19).

The observation that B cells, but not educated T cells, are radiosensitive is ancillary to Hartmann's (28) finding that the PFC produced are derived from precursors in the B cell population, based on the demonstration that they are of the isoantigenic type of the B cell population and not the T cell population.

The radioresistance of educated T cells is worthy of comment. This observation differs from the reported radiosensitivity of native thymus cells (1) and educated T cells (7) assayed *in vivo*. It should be emphasized that the radioresistance of the *function* of the educated T cell *population* has been observed. In this light, a number of explanations may be presented. Educated T cell radioresistance could be explained on the basis of the number of antigen-specific T cells produced during our education procedure. Education results in an increase in antigen-specific T cells (6) and it is possible that the specific population was expanded to the extent that a significant number of specific T cells could escape inactivation by exposure to 2000 R. Other possibilities are that educated T cells have a pre-formed product required for cooperation with B cells or that they do not have to divide to function in the *in vitro* cultures. Since our observation closely parallels the reported radioresistance of carrier primed lymphoid cells (17, 24), further experimentation with both experimental systems is required to determine the basis of the radioresistance of the cooperative function.

Acknowledgment. We thank Professor Richard W. Dutton for his generous support and advice. D.C.V. was the recipient of a grant from the University of California Cancer Research Coordinating Committee. We thank I. Goldman and the Salk Institute for Biological Studies for the use of their radiation facilities.

REFERENCES

1. Claman, H. N., Chaperon, E. A. and Triplett, J. L., J. Immunol., *97:* 828, 1966.
2. Mitchell, G. F. and Miller, J. F. A. P., Proc. Nat. Acad. Sci., *59:* 296, 1968.
3. Davies, A. J. S., Leuchars, E., Wallis, V., Marchant, R. and Elliot, E. V., Transplantation, *5:* 222, 1967.
4. Mitchell, G. F. and Miller, J. F. A. P., J. Exp. Med., *128:* 821, 1968.
5. Nossal, G. J. V., Cunningham, A., Mitchell, G. F. and Miller, J. F. A. P., J. Exp. Med., *128:* 839, 1968.
6. Shearer, G. M. and Cudkowicz, G., J. Exp. Med., *130:* 1243, 1969.
7. Claman, H. N. and Chaperon, E. A., Transplant. Rev., *1:* 92, 1969.
8. Mosier, D., Science, *158:* 1573, 1967.
9. Pierce, C. W., J. Exp. Med., *130:* 345, 1969.
10. Mishell, R. I., Dutton, R. W. and Raidt, D. J., Cell. Immunol., *1:* 175, 1970.
11. Dutton, R. W., McCarthy, M. M., Mishell, R. I. and Raidt, D. J., Cell. Immunol., *1:* 196, 1970.
12. Hoffmann, M. and Dutton, R. W., Science, *172:* 1047, 1971.
13. Bosma, M. J., Perkins, E. H. and Makinodan, T., J. Immunol., *101:* 963, 1968.
14. Mosier, D. E. and Coppleson, L. W., Proc. Nat. Acad. Sci., *61:* 542, 1968.
15. Groves, D. L., Lever, W. E. and Makinodan, T., J. Immunol., *104:* 148, 1970.
16. Rajewsky, K., Schirrmacher, V., Nase, S. and Jerne, N. K., J. Exp. Med., *129:* 1131 1969.
17. Katz, D. H., Paul, W. E., Goidl, E. A. and Benacerraf, B., J. Exp. Med., *132:* 261, 1970.
18. Raff, M. C., Nature, *226:* 1257, 1970.
19. Dutton, R. W., Campbell, P., Chan, E., Hirst, J., Hoffmann, M., Kettman, J., Lesley, J., McCarthy, M., Mishell, R. I., Raidt, D. J. and Vann, D. In *Second Int. Convoc. Immunol.: Cellular Interactions in the Immune Response*, Edited by R. T. McClusky, S. Cohen, G. Cudkowicz and J. Mohn, S. Karger, New York, In press.
20. Mitchison, N. A., Rajewsky, K. and Taylor, R. B., in *Developmental Aspects of Antibody Formation and Structure*, Edited by J. Sterzl and M. Riha, Vol. 2, p. 547, Academic Press, New York, 1970.
21. Munro, A. and Hunter, P., Nature, *225:* 277, 1970.
22. Hirst, J. A. and Dutton, R. W., Cell. Immunol., *1:* 190, 1970.
23. Doria, G., Martinozzi, M., Agarossi, G. and Di Pietro, S., Experientia, *25:* 410, 1970.
24. Kettman, J. and Dutton, R. W., Proc. Nat. Acad. Sci., *68:* 699, 1971.
25. Trowbridge, I. S., Lennox, E. S. and Porter, R. R., Nature, *228:* 1087, 1970.
26. Lesley, J. F., Kettman, J. R. and Dutton, R. W., J. Exp. Med., In press.
27. Hartmann, K-U, Behringwerk-mitteilungen, *49:* 208, 1969.
28. Hartmann, K-U, J. Exp. Med., *132:* 1267, 1970.
29. Rittenberg, M. B. and Pratt, K. L., Proc. Soc. Exp. Biol. Med., *132:* 575, 1969.
30. Mishell, R. I. and Dutton, R. W., J. Exp. Med., *126:* 423, 1967.
31. Kettman, J. and Dutton, R. W., J. Immunol. *104:* 1558, 1970.
32. Makinodan, T., Sado, T., Groves, D. L. and Price, G., Curr. Topics Microbiol. Immunol., *49:* 81, 1969.
33. Roseman, J., Science, *165:* 1125, 1969.
34. Talmage, D. W., Radovich, J. and Hemmingsen, H., J. Allergy, *43:* 323, 1969.
35. Basten, A., Miller, J. F. A. P., Warner, N. L. and Pye, J., Nature, *231:* 104, 1971.
36. Chan, E. L., Mishell, R. I. and Mitchell, G. F., Science, *170:* 1215, 1970.

RESPONSE OF MOUSE T AND B LYMPHOCYTES

TO SHEEP ERYTHROCYTES

J.H.L. Playfair

THE primary immune response of mice to sheep red blood cells (SRBC) involves two types of lymphocytes: the antibody-forming cell series, whose precursors are found in bone marrow (B cells), and cooperating or "helper" cells of uncertain function whose precursors are found in the thymus (T cells)[1]. Within 24 h of an intravenous injection of SRBC, an increase of haemolytic antibody plaque-forming cells (p.f.c.) occurs in the spleen, reaching a peak 5 days later. Part, but not all, of this increase is due to division among the B cells[2]. T cells in the spleen also undergo a wave of mitosis, detectable from the second to the fifth day[3].

Cooperation experiments *in vivo* in which bone marrow was used as the source of B cells and thymus as the source of T cells[4] have shown that the T cells must be intact and viable, their cooperating function being destroyed by prior X-irradiation with 900 rad[5]. One explanation would be that the T cell mitosis mentioned above is necessary for cooperation, and is preventable by irradiation.

In another cooperation system, however, where marrow-derived (B) cells were cultured *in vitro* with spleen cells previously primed to a carrier, and the antibody response to a hapten measured, it was found that the cooperating function of the carrier-primed spleen cells could be destroyed by anti-θ (that is, anti-T cell) serum but not by 4,000 rad of X-rays[6]. In another similar model, in which carrier-primed guinea-pig spleen cells were transferred to syngeneic animals, the primed cells cooperated in an anti-hapten response despite 5,000 rad of X-rays, but were unable to synthesize DNA in response to the carrier antigen[7]. The implication is that spleen T cells which have already been primed, and presumably divided, can no longer be prevented from cooperating by irradiation. Therefore the change from radiosensitivity to radioresistance may mark one stage in the response of T cells to antigen.

This suggests a functional assay for such a response: T cells primed by SRBC should survive irradiation and be available to cooperate with subsequently injected bone marrow (B) cells given with a second dose of SRBC. That is to say, primed irradiated mice given marrow and SRBC should make more p.f.c. than if either the priming or the bone marrow were omitted. Such an effect has in fact been observed by Zaretskaya et al.[8]. By analogous reasoning, if any B cells acquire radioresistance after priming, it should be possible to reveal them by the injection of thymus (T) cells and SRBC.

In the experiments reported here, (NZB × Balb/c)F1 or (C57Bl × Balb/c)F1 mice were primed with different numbers of SRBC intravenously, and at various intervals thereafter given 850 rad of 230 kVp X-rays followed by 2×10^7 SRBC mixed with either 15×10^6 syngeneic marrow or 5×10^7 syngeneic thymus cells, the direct (IgM) p.f.c. in their spleens being counted 8 days later, by a standard technique[9]. Two controls were necessary. Some mice were primed but received no injection after irradiation; p.f.c. in their spleens 8 days later were assumed to be survivors from the interrupted primary response. Other mice were not primed but received marrow (or thymus) and SRBC; their p.f.c. are probably produced by unaided B cells[10], or by radioresistant cells present in the unprimed spleen. Both these control values were subtracted from the experimental p.f.c. counts, and the remaining p.f.c. taken to be the fruit of cooperation.

Table 1 shows the results of an experiment in which (NZB × Balb/c)F1 mice were primed with 2×10^7 SRBC 2 days before irradiation. Mice then given bone marrow and SRBC made

Table 1 Effect of Priming with SRBC on Normal and Thymus-deprived Mice

	Unprimed mice Marrow+ SRBC	No cells	Primed mice Marrow+ SRBC	Coopera- tive p.f.c.
Normal	1,229 (914–1,651)	4,978 (4,244–5,839)	13,254 (11,793–14,896)	7,047
Thymx. + ALG × 2	997 (679–1,465)	2,876 (1,775–4,660)	4,664 (3,211–6,776)	791

Eight-day spleen p.f.c. in irradiated normal or thymus-deprived (NZB × Balb/c)F1 mice injected with 15×10^6 marrow cells and 2×10^7 SRBC, some of the mice being primed with 2×10^7 SRBC 2 days before irradiation. The numbers are the mean ± one standard error of the log-converted data.

7,000 more p.f.c. than would be predicted by adding together the two control values. This cooperative p.f.c. response, however, was substantially reduced in mice deprived of many of their T cells by thymectomy and two injections of antilymphocyte globulin. Thus the expectation was fulfilled that this method can be used to detect T cells responding to the initial SRBC injection.

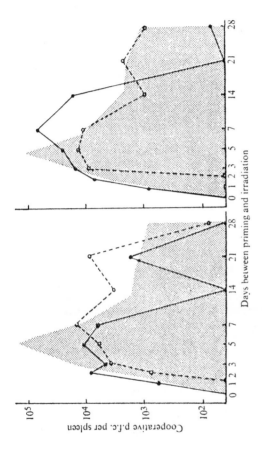

Fig. 1 Cooperative spleen p.f.c. responses 8 days after irradiation in (C57Bl × Balb/c)F1 (left) and (NZB × Balb/c)F1 (right) mice primed with 2×10^7 SRBC at various times before irradiation. "T cells" (●) denotes the corrected p.f.c. obtained when marrow cells and SRBC were injected; "B cells" (○) those obtained when thymus cells and SRBC were injected. The shaded area represents the p.f.c. in the spleen at the time of irradiation—that is, the normal primary response.

91

Table 2 shows the way in which the effect varies with the interval between priming and irradiation. The cooperative p.f.c. response becomes positive the day after priming, reaches a peak on day 7, and has subsided by the third week. Kettman and Dutton, with their *in vitro* assay, found very similar kinetics[6]. By contrast, a corresponding effect in the B cells (measured by the addition of thymus cells and SRBC) is not detectable until the third day, but has still not subsided 4 weeks later.

Table 2 Effect of Varying the Interval between Priming with SRBC and Irradiation

Day	No cells	Marrow + SRBC	Cooper-ative p.f.c.	Thymus + SRBC	Cooper-ative p.f.c.	SRBC alone	Cooper-ative p.f.c.
0		1,229	0	237	0		
−1	5,240	2,719	815	342	0	675 *	0
−2	6,190	13,254	7,047	4,261	0	4,978 *	0
−3	684	16,801	14,788	9,801	8,880	3,287	2,565
−5	4,261	31,268	25,418	18,206	13,348	7,149	2,490
−7	2,762	73,042	69,051	13,923	10,924	30,023	27,223
−14	66	17,806	16,509	1,235	932	2,709	2,605
−21	297	1,110	0	2,784	2,250	191	0
−28	329	1,628	70	1,538	972	170	0

* Used as controls; see text.
Eight-day spleen p.f.c. in irradiated $(NZB \times Balb/c)F1$ mice primed with 2×10^7 SRBC at various times before irradiation and injection of 15×10^6 marrow or 5×10^7 thymus cells and/or 2×10^7 SRBC. The numbers are geometric means. The correction procedures are described in the text.

The last column in Table 2 shows the result of injecting only SRBC after irradiation. As might be predicted, when both T and B cells are in the radioresistant state—that is, between the third and the fourteenth day—a cooperative response is obtained without the addition of either B or T cells. On the other hand, with priming 1 or 2 days before irradiation, the injection of SRBC alone caused a reduction of "surviving" p.f.c., presumably by some feedback mechanism. In such cases it seemed more appropriate to use the reduced counts as the "survivor" controls.

In Fig. 1 are plotted the cooperative p.f.c. responses for mice of both strains primed with 2×10^7 SRBC, and Fig. 2 shows the 7-day responses for various other priming doses of SRBC. T cell priming precedes B cell priming by about one day (Fig. 1). This agrees with the results of Davies *et al.*[3] who measured the mitotic wave in T and B cells using chromosome markers. Since the primary p.f.c. response in the intact mouse starts the day after giving SRBC, it is apparently possible for B cells to make antibody before entering the radioresistant state. The rise of T cell activity could be explained by T cell proliferation or recruitment from outside the spleen, and its rather rapid subsidence might likewise be due to the primed cells dying, leaving the spleen, or passing once more into a radiosensitive phase.

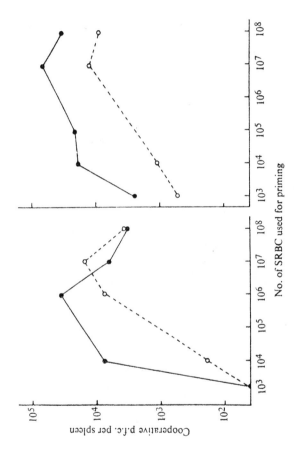

Fig. 2 Cooperative p.f.c. responses in (C57Bl × Balb/c)F1 (left) and (NZB × Balb/c)F1 (right) mice primed with various numbers of SRBC 7 days before irradiation. Other details as in Fig. 1. ●, T cells; ○, B cells.

Low doses of SRBC prime T cells better than they prime B cells (Fig. 2). One cannot, of course, compare the respective p.f.c. counts directly, because they are partly determined by the number of marrow and thymus cells injected. But 2×10^4 SRBC seems to be a more effective dose for priming T than B

Table 3 Priming with other RBC; Comparison of "Cross-priming" and Cross-reaction

RBC injected	No cells	Marrow + SRBC	Cooperative p.f.c.	Peak (5-day) SRBC p.f.c. in unirradiated mice
Day-2				
Rabbit	92	6,976	5,747	269
Pig	123	6,035	4,806	1,253
Human	39	4,023	2,794	—
Horse	26	2,275	1,046	395
Chicken	38	1,514	285	711
Day-7				
Rabbit	175	3,451	2,047	
Horse	5,014	37,214	30,009	
Chicken	108	1,874	537	

Eight-day spleen SRBC p.f.c. in irradiated (NZB × Balb/c)F1 mice primed with 2×10^7 RBC of various kinds 2 or 7 days before irradiation and injection of 15×10^6 marrow cells and 2×10^7 SRBC. Details as in Fig. 2.

cells, and it also gives only 2% of the maximal p.f.c. response in intact mice. It may be that the property detected by the assay for radioresistance in the T cells is related not so much to the kinetics of the primary response itself as to the development of memory. In fact the time courses for B and T cell priming shown here are closely similar to those for B and T cell memory obtained by Cunningham and Sercarz[11], using a system in which different RBC determinants acted as "carrier" and "hapten". It is also notable that there is a consistent strain difference, the NZB hybrid responding more vigorously to priming at the level of both the T and the B cells.

One obvious application of this kind of system is in the study of T cell specificity, since one can use other antigens to prime the T cells, and then test them in a cooperative response with marrow and SRBC. Kettman and Dutton[6] and Mitchell and Miller[12], using a cell-transfer system with primed thymus-derived cells, have shown that horse and chicken RBC did not prime the T cells for an SRBC response. On the other hand Hartmann has reported an effect with horse RBC[13].

Table 3 shows some preliminary results using five kinds of RBC, which vary greatly in their ability to prime the T cells for an SRBC response. Once again, the NZB hybrid was much more susceptible to this "cross-priming". Curiously enough, the most effective RBC (pig and rabbit) did not apparently cross-react much more with SRBC than the least effective (chicken), as judged by the SRBC p.f.c. they induced in intact mice. This raises the question whether T cells may recognize antigens as cross-reacting when the B cells do not, which in turn suggests that the antigen receptor on the T and the B cells might be different. Alternatively, part of the cooperative effect, especially in the NZB hybrid, might be due to a non-antigen-specific factor, liberated by specifically stimulated T cells but active on all B cells, or on other T cells. A full account of these and other studies on T and B cell specificity will be published elsewhere.

[1] Playfair, J. H. L., *Clin. Exp. Immunol.*, **8**, 839 (1971).
[2] Tannenberg, W. J. K., and Malaviya, A. N., *J. Exp. Med.*, **128**, 895 (1968).
[3] Davies, A. J. S., Leuchars, E., Wallis, V., and Koller, P. C., *Transplantation*, **4**, 438 (1966).
[4] Claman, H. N., Chaperon, E. A., and Triplett, R. F., *Proc. Soc. Exp. Biol. Med.*, **122**, 1167 (1966).
[5] Claman, H. N., Chaperon, E. A., and Selner, J. C., *Proc. Soc. Exp. Biol. Med.*, **127**, 462 (1968).
[6] Kettman, J., and Dutton, R. W., *Proc. US Nat. Acad. Sci.*, **68**, 699 (1971).
[7] Katz, D. H., Paul, W. E., Goidl, E. A., and Benacerraf. B., *Science*, **170**, 462 (1970).
[8] Zaretskaya, Y. M., Panteleev, E. I., and Petrov, R. V., *Nature*, **221**, 567 (1969).
[9] Playfair, J. H. L., *Immunology*, **15**, 815 (1968).
[10] Playfair, J. H. L., and Purves, E. C., *Immunology*, **21**, 113 (1971).
[11] Cunningham, A. J., and Sercarz, E. E., *Europ. J. Immunol.* (in the press).
[12] Mitchell, G. F., and Miller, J. F. A. P., *Proc. US Nat. Acad. Sci.*, **59**, 296 (1968).
[13] Hartmann, K. U., *J. Exp. Med.*, **132**, 1267 (1970).

Cell Cooperation in Cellular Immune Responses

COOPERATION OF LYMPHOID CELLS IN AN IN VITRO GRAFT REACTION SYSTEM

II. The "Bone Marrow-Derived" Cell[1]

Peter Lonai and Michael Feldman

SUMMARY

An in vitro graft reaction system in which rat lymphocytes were sensitized in culture against mouse fibroblasts, then were tested for their capacity to effect immune lysis of mouse cells, was found to be based on a bicellular lymphoid interaction. Rat thymus cells, which by themselves are incapable of effecting the lytic reaction following sensitization on mouse fibroblasts, when combined with spleen cells, confer specific lytic activity on the mixed cell population. Experiments were made to define the tissue origin and properties of the spleen cells which are capable of interacting with the thymus cells. A mixed population of bone marrow and thymus cells, combined and cultured on mouse cells, did not manifest immune lysis of mouse target cells. Mixed populations consisting of spleen cells of thymectomized, X-irradiated mice inoculated with rat marrow and of rat thymus cells combined and cultured on mouse fibroblasts did show lytic activity. Thus, bone marrow-derived cells within the spleen interact with thymus cells to effect lysis of target monolayers. The capacity of such bone marrow cells to recognize transplantation antigens was studied. Rat bone marrow-derived cells (following colonization of B6D2F1 mouse spleens), when admixed with rat thymus cells, could be sensitized and would effect lytic activity on C3H, but not on C57BL and DBA/2 strain, fibroblasts. Thus, the rat bone marrow-derived cells seemed to recognize the recipient's mouse antigen, since they manifested unresponsiveness toward fibroblasts of similar antigenic constitution. We conclude that: (1) the in vitro graft reaction system is based on interaction between thymus-processed and nonthymus-processed lymphocytes; (2) the nonthymus-processed cells derive from the bone marrow but are not identical with the original bone marrow cells. They seem to represent a differentiation product of bone marrow cells; (3) both the thymus- and the nonthymus-processed cells are capable of recognizing the mouse antigens. The question of which is the killer cell remains open. The necessary participation of thymus cells in the killing process is discussed.

The induction of antibody production was observed to be based on a bicellular interaction of lymphocytes (8, 10, 25, 39). The two cells cooperating in humoral immune responses were demonstrated to have different ontogenic and functional properties. Although, most probably, both originate in the bone marrow (11, 14, 15, 27, 42, 43), one undergoes "processing" within the thymus prior to its appearance in the peripheral lymphoid organs (11, 16). The thymus-processed cells function as antigen-recognizing cells and are themselves incapable of effective antibody production (11, 25, 26, 35, 37). The actual production of antibody is performed by bone marrow-derived cells not processed by the thymus (28, 30) which interact, apparently via the antigen, with the thymus-derived cells (29, 35, 36).

A bicellular mechanism was claimed to be operative in cell-mediated immune reactions as well. This mechanism for allograft reactions was found by Argyris (2). In graft-versus-host (GVH) reactions, thymus-marrow interaction was also demonstrated (4, 7, 17), although not unequivocally (12, 38, 41).

[1] This work was supported by a grant from the Max and Ida Hillson Foundation, New York, and by a grant from the Freudenberg Foundation for Research on Multiple Sclerosis, Weinheim, Germany.

98

In order to analyze whether a bicellular mechanism operates in cell-mediated immune reactions, we applied the in vitro heterograft system developed in our laboratory (5, 16, 22). In this system, rat lymphoid cells are cultured and thus are sensitized on mouse fibroblast monolayers, where they transform, multiply, and become capable of destroying the mouse cells. Cytolysis, found to be antigenically triggered, is measured by the radiochromium release assay (5). We found that lymphocytes of different lymphoid tissues have different capacities to effect lysis. Thus, thymus cells almost completely lack activity. However, when thymus cells were combined with rat spleen cells, a significant synergistic effect was observed. We found that the antigenic specificity of the reaction was determined by the antigen which had sensitized the thymus cells, but the lytic process per se took place only in the presence of spleen cells (22). These studies suggested that, in cellular immune reactions too, the thymus cell is the antigen-reactive cell capable of initiating the response. Since, in our studies, the interacting partner of thymus cells was spleen cell suspensions, we aimed the present study to define the nature and tissue origin of this interacting cell.

MATERIALS AND METHODS

Animals. Inbred Lewis female rats, 2½–4 months of age (Microbiological Associates; Weizmann Institute Breeding Center) were used as donors of thymus, spleen, and bone marrow cells.

Embryos of C3HeB, C57BL/6, and DBA/2 mice from our breeding center were used as the sources of fibroblast monolayers. Female (C57BL/6 × DBA/2)F$_1$ (B6D2F1) hybrid mice (The Jackson Laboratory) were used as hosts for rat bone marrow cells in the heterochimera experiments. The spleen and lymph nodes of these animals were used.

Tissue culture techniques. Lymphocyte suspensions from the described sources or their combinations were cultured in Eagle's minimum essential medium plus 20% horse serum at 37 C with 7% CO_2 gas supply. The preparation of cell suspensions and the detailed tissue culture conditions were described previously (5, 22).

The assay systems. Two assay systems were employed to test the capability of various cell populations to cooperate with rat thymus cells

in the lytic reaction directed against mouse fibroblasts.

In the *first system,* freshly prepared thymus cell suspensions were combined with the cell suspension to be tested in different ratios, so that the final quantity of cells in one Petri dish was 20×10^6. The cells were cultured on a mouse fibroblast sensitizing monolayer for 5 days. After this, the cells were collected, and aliquots of 2.5 $\times 10^6$ viable cells (as demonstrated by dye exclusion test) were transferred to ^{51}Cr-labeled test monolayers for 20 hr for estimation of chromium release. The level of chromium release expressed as percentage of radioactivity compared to the total radioactivity of the culture is the measure of the cell-mediated damage (5, 22).

The *second assay system* was based on our former studies on synergism between in vitro sensitized rat thymus cells and nonsensitized rat spleen cells (22). For sensitization, rat thymus cells were plated on mouse fibroblast monolayer cultures in 60-mm Falcon Petri dishes at doses of $40–50 \times 10^6$ cells/dish and were cultured for 8–11 days. After this period, the cells were collected, counted, and mixed in different ratios with the cell suspension to be tested. Since mixtures of in vitro sensitized thymus cells and nonsensitized spleen cells lyse the adequate target monolayer immediately following explantation, with a peak of activity on the 2nd day (22), the mixtures were plated (for 20 hr) either directly on chromiated target monolayers or, after 2 days of further incubation, on intermediate, nonlabeled cultures. In both cases, 2.5 $\times 10^6$ viable nucleated cells were transferred to the labeled monolayers. The chromium release caused by the constituents of the cell combinations alone also was measured. Enhancement attributable to synergism in the mixed population was expressed as percentage of increase over the chromium release caused by the more active component alone. Significance was calculated according to Student's *t* test. Details of both assay systems were described previously (22).

Production of heterochimeras. B6D2F1/J female mice, aged 12 weeks, were thymectomized and, 7–10 days later, exposed to 950–1,000 R total body X-irradiation in a Perspex box (containing an inner Radocon dosimeter, using a Picker-Vanguard X-ray machine at 250 kv, 15 ma, 60 R/min; FSD, 50 cm; filters, 0.5 mm Cu, 1.0 mm Al). After irradiation, two groups of animals were injected i.v. with 40 or 90×10^6

TABLE 1. Activity of cultures consisting of nonsensitized thymus and bone marrow cells[a]

Cells comprising culture[b] (ratio)	Lysis $\pm$ SD	Lytic increase (%)	Significance
Thymus	8.8		
Spleen	15.3 $\pm$ 0.9		
Bone marrow	2.5 $\pm$ 0.2		
1. Thymus + spleen			
1:3	22.7 $\pm$ 0.9	32.6	$P \ll 0.001$
1:2	28.0 $\pm$ 1.8	45.4	$P < 0.001$
1:1	26.1 $\pm$ 0.9	41.4	$P \ll 0.001$
2:1	30.9 $\pm$ 1.0	50.5	$P < 0.001$
3:1	34.4 $\pm$ 0.8	55.6	$P < 0.001$
2. Thymus + bone marrow			
1:3	1.4	0.0	—
1:2	2.0 $\pm$ 0.1	0.0	—
1:1	1.6 $\pm$ 0.3	0.0	—
2:1	2.3 $\pm$ 1.4	0.0	—
3:1	2.9	0.0	—

[a] Rat thymus, spleen, and bone marrow cell suspensions and combinations of thymus and spleen and of thymus and bone marrow cells were plated on C3H mouse monolayers. The cells were cultured for 5 days, after which they were transferred in doses of 2.5×10^6 cells to ^{51}Cr-labeled C3H cultures for the measurement of chromium release.

[b] Part 1 of this experiment was published previously (22) without part 2. The two experiments were made with the same materials.

freshly prepared rat bone marrow cells, respectively. From the 3rd day prior to irradiation to the 14th day after exposure, the animals received acidified drinking water and 0.1 ml of Pyopen i.m. (Beecham Research Laboratories, Brentford, England) every other day.

Test for chimeric state. The alkaline phosphatase method of Gömöri (19) was used. This test is based on the fact that rat granulocytes are alkaline phosphatase-positive, whereas mouse granulocytes are not. Peripheral blood samples from each mouse were tested 16 days after inoculation with rat bone marrow. The granulocytes were counted and the percentage of alkaline phosphatase-positive rat granulocytes was determined.

RESULTS

Activity of cultures consisting of bone marrow and thymus cells. A direct approach to defining the partner of the thymus cells in the cellular interaction which takes place in the in vitro graft reaction system was to test the capacity of rat bone marrow cells to produce cytolytic activity when admixed with thymus cells. Therefore, fresh thymus, spleen, and bone marrow cell suspensions were prepared for culturing on C3H monolayers.

The following experimental groups were formed: thymus, spleen, or bone marrow cells alone, and combined populations of thymus and spleen and of thymus and bone marrow cells, at different ratios. The cells were cultured for 5 days on sensitizing monolayers, after which they were transferred to labeled test monolayers for determination of lytic activity. As a result (Table 1), cell suspensions consisting of thymus and spleen cells gave a significantly enhanced lytic activity over that of spleen cells. On the other hand, bone marrow cell suspensions revealed no lytic activity, whether cultured alone or in combination with thymus cells. The experiment with fresh bone marrow and thymus cells was repeated twice with the same result. Rat spleen cells could produce a significant enhancement of the lytic activity when combined with thymus cells, but bone marrow cells apparently lacked this capacity.

In our previous studies, we demonstrated that presensitized thymus cells combined with fresh rat spleen cells give a more extensive enhancement of lytic activity than do combinations of fresh thymus and spleen cells (22). Therefore, in the second series of experiments, bone marrow cells were mixed with thymus cells presensitized on C3H monolayers for 8 days. The cells were transferred to intermediate monolayers for 2 days, after which the lytic activity was tested on labeled test monolayers. The results are recorded in Table 2. Experiments 1 and 2 demonstrate that bone marrow cells were incapable of interacting effectively in this combination, as well. Under similar conditions, spleen cells together with sensitized thymus cells gave a significantly enhanced reaction (Table 2, experiment 3), as described previously (22). These experiments were repeated 3 times, and we constantly found that rat bone marrow cells failed to manifest

TABLE 2. *Activity of cultures consisting of sensitized thymus and fresh bone marrow cells[a]*

Experiment No. and cells comprising culture[b] (ratio)	Lysis ± SD	Lytic increase (%)	Significance[c]
1. T°	4.0 ± 0.22		
BM	1.2 ± 0.19		
T° + BM			
3:1	4.8 ± 0.80	16.5	NSD
2. T°	5.1 ± 1.6		
BM	0.5 ± 0.2		
T° + BM			
3:1	6.9 ± 2.5	26.0	NSD
1:1	3.0 ± 0.8	—	—
1:3	1.3 ± 0.9	—	—
3. T°	4.8 ± 0.3		
S^L	2.1 ± 1.5		
T° + S^L	30.2 ± 0.8	84.1	$P < 0.001$

[a] Rat thymus cells were cultured on C3H fibroblast monolayers for 8 days. After this, they were collected and mixed with either rat bone marrow or spleen cells. Thymus, bone marrow, and spleen cells and their appropriate mixtures were transferred in doses of 8×10^6 cells to nonlabeled intermediate monolayers. Two days later, they were transferred in doses of 2.5×10^6 cells to ^{51}Cr-labeled C3H test monolayers for the measurement of chromium release.

[b] T° = sensitized thymus cells; BM = rat bone marrow cells; S^L = rat spleen cells.

[c] NSD = no significant difference.

cytolytic activity, either alone or when combined with thymus cells.

The production and testing of rat "bone marrow-derived" cells. It is conceivable that spleen and lymph nodes contain, besides thymus-derived cells, nonthymus-processed lymphocytes, which derive from the bone marrow, but which might have undergone further differentiation in the peripheral lymphoid organs. To produce cell populations containing such cells and to test their capacity to cooperate with thymus cells, the following experimental design was applied. Lethally irradiated mice were repopulated with rat bone marrow cells. Spleen cell suspensions of these heterochimeras were mixed with syngeneic rat thymus cells. In most experiments, the thymus cells were presensitized on mouse monolayers. The mixtures of presensitized thymus and heterochimeric spleen cells were tested for lytic activity on mouse fibroblast cultures genetically similar to those used for the sensitization of the thymus cells. The scheme of these experiments is shown in Figure 1.

The characteristics of the rat-mouse heterochimeras and of the cultures made from their spleens. About 40–50% of the heterochimeras survived more than 2–3 weeks postirradiation. There were no subsequent losses until the 60th–75th days, when the experiments were terminated. The chimeric state of the animals was tested. Alkaline phosphatase-positive rat granu-

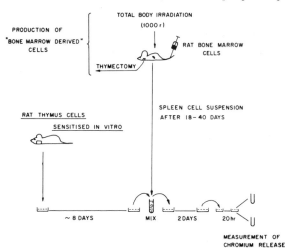

FIGURE 1. The scheme of experiments with chimera cells. See explanation in text.

TABLE 3. The activity of lymphoid cells from heterochimeric mice[a]

Source of spleen cells	Resulting lysis ± SD in experiment No.		
	1	2	3
Heterochimeric mice	0.2 ± 0.2	1.7	0.0
Rats (controls)	20.6 ± 0.47	35.1 ± 0.62	35.4 ± 1.65

[a] Cell suspensions were made from spleens of thymectomized, irradiated B6D2F1 mice which were injected with Lewis rat bone marrow and spleen cells. The cells were cultured for 5 days on C3H mouse fibroblast cultures, after which the chromium release was determined by the transfer of 2.5×10^6 cells to ^{51}Cr-labeled test monolayers.

locytes were found in the peripheral blood of all treated animals.

The histological picture of the spleen and lymph nodes was characterized by a decrease in small lymphocytes, mainly in the so called thymus-dependent areas (32). Germinal centers were absent or decreased in number. Medium and large lymphocytes were predominant in the spleen cell suspensions.

When heterochimeric spleen cell suspensions were cultured on fibroblast monolayers, we regularly observed that, in the cell population, which originally contained mostly lymphoid cells, the number of macrophage-like cells increased with time. After 4 days in culture, the macrophages became predominant. We tested the lytic activity of heterochimeric spleen cells by culturing them on C3H mouse monolayers for 5 days, then transferring them to labeled C3H test monolayers. Rat spleen cells were processed as controls. The heterochimeric spleen cells revealed no lytic activity. The results of three such experiments are shown in Table 3.

Activity of cultures consisting of rat thymus and chimeric spleen cells. Cell suspensions were made from spleen cells of heterochimeric mice and from the thymus and spleen cells of Lewis rats. These suspensions and the combinations of thymus plus chimeric cells or thymus plus rat spleen cells were plated on C3H mouse fibroblast monolayers. The cells were incubated for 5 days,

then were transferred to labeled test monolayers for the measurement of chromium release. The results of two such experiments are recorded in Table 4. The combinations of thymus and rat spleen cells revealed synergistic activity, as described previously (22), whereas the combinations with chimeric spleen cells produced no activity. We made two more such experiments, and both gave the same results.

In these experiments, as mentioned, large quantities of macrophages were observed. They were present in all cultures containing heterochimeric spleen cells, and their amounts increased with time. The presence of macrophages could account for the inactivity of these cultures. We found (Lonai and Feldman, unpublished data) that more than 5% added macrophages had a

TABLE 4. Activity of cultures consisting of non-sensitized rat thymus cells and chimera spleen cells[a]

Experiment No. and cells comprising culture[b] (ratio)	Lysis ± SD	Lytic increase (%)	Significance
1. T	3.9		
S^H	0.0		
S^L	35.4 ± 1.7		
T + S^H			
3:1	4.1	0.0	
T + S^L			
3:1	46.6 ± 0.73	24.0	P < 0.05
2. T	7.3 ± 0.6		
S^H_1	0.0 ± 0.2		
S^H_2	0.0 ± 0.5		
S^L	11.1 ± 0.3		
T + S^L			
3:1	15.4 ± 0.1	28.0	P < 0.001
T + S^H_1			
3:1	1.0 ± 1.0	0.0	
1:1	1.0	0.0	
T + S^H_2			
3:1	7.0 ± 0.6	0.0	
1:1	2.5 ± 0.3	0.0	

[a] Cell suspensions were made from rat thymuses, rat spleens, and from heterochimeric mouse spleens. Combinations of these, and the suspensions alone, were cultured on C3H fibroblast monolayers for 5 days, after which cells were transferred to ^{51}Cr-labeled test monolayers for the measurement of chromium release.
[b] T = Lewis rat thymus cells; S^H = heterochimeric spleen cells (pooled from 3 animals); S^H_1 and S^H_2 = heterochimeric spleen cells (two separate animals); S^L = Lewis rat spleen cells.

102

TABLE 5. Cooperation between sensitized rat thymus cells and lymphoid cells of rat-mouse heterochimeras[a]

Experiment No. and cells comprising culture[b] (ratio)	Lysis ± SD	Lytic increase (%)	Significance
1. T°	11.4 ± 1.9		
S^H	0.0		
T° + S^H			
3:1	19.7 ± 0.7	42	$P < 0.02$
2. T°	11.0 ± 0.3		
S^H	1.5 ± 0.2		
L^H	0.0		
T° + S^H			
3:1	18.1 ± 0.6	39	$P < 0.001$
1:1	14.5 ± 0.3	24	$P < 0.05$
T° + L^H			
3:1	17.2 ± 0.4	37	$P < 0.001$

[a] Rat thymus cells were cultured on C3H fibroblast monolayers (for 8 days in experiment 1; for 10 days in experiment 2). After this, the sensitized thymus cells were admixed with heterochimeric spleen or lymph node cells. Both the mixtures and their components alone were transferred to intermediate monolayers for 2 days, after which the chromium release was measured by the transfer of 2.5 × 10[6] cells to ^{51}Cr-labeled C3H test monolayers.
[b] T° = sensitized thymus cells; S^H = heterochimeric spleen cells; L^H = heterochimeric lymph node cells.

strong inhibitory effect on the cytolytic activity of our cultures. This observation is similar to the findings of Perkins and Makinodan (33) and of Parkhouse and Dutton (31) on the effect of added macrophages on humoral immune reactions.

Activity of cultures consisting of sensitized rat thymus cells and chimeric spleen cells. In this series of experiments, chimeric spleen cells were combined with thymus cells sensitized on mouse fibroblasts (see Fig. 1). The combinations were cultured on an intermediate monolayer for a short time (less than that needed for the appearance of macrophages), after which they were transferred to test monolayers. The following experiments (Table 5, experiments 1 and 2) were made. Thymus cells were sensitized on C3H monolayers. The following groups were formed in the two experiments: sensitized thymus cells alone and mixtures of sensitized thymus plus spleen or lymph node cells from the chimeric mice, respectively. After 2 days of cul-

ture on intermediate monolayers, the cells were transferred to labeled test monolayers, and the chromium release was measured. Both spleen and lymph node cells of the chimeric mice, when mixed with sensitized thymus cells, gave chromium release values exceeding those of the sensitized thymus or chimeric cells alone, or their sum (Table 5).

In the next series of experiments, we tested whether the increased lytic activity observed with cells from the chimeric mice is indeed the result of rat cells present in the chimeric spleens or whether it is attributable to the mouse spleen cells which were obviously admixed with rat cells in the host. Experiments were made, similar to those described in Table 5, but additional groups were formed by mixing sensitized rat thymus cells with spleen cells of untreated B6D2F1 mice. Table 6 records the results. Combinations with chimeric spleen cells gave significantly increased chromium release whereas, in groups containing normal mouse spleen cells, no activity could be measured. Thus we demon-

TABLE 6. The activity of cultures consisting of sensitized rat thymus cells and Lewis-B6D2F1 heterochimera cells, and that of sensitized rat thymus cells and B6D2F1 mouse spleen cells[c]

Cells comprising culture[b] (ratio)	Lysis ± SD	Lytic increase (%)	Significance
T°	8.4 ± 0.9		
S^H	0.0		
T° + S^H			
3:1	13.7 ± 0.8	39	$P < 0.001$
1:1	16.9 ± 2.0	50	$P < 0.001$
1:3	26.1 ± 3.3	68	$P < 0.001$
S^B	0.2 ± 0.1		
T° + S^B			
3:1	6.0 ± 0.5	0.0	—
1:1	4.6 ± 0.6	0.0	—
1:3	3.9 ± 0.8	0.0	—

[a] Rat thymus cells were cultured on C3H fibroblast monolayers for 8 days. Spleen cells of thymectomized, irradiated B6D2F1 mice, repopulated with rat bone marrow cells or spleen cells of untreated B6D2F1 mice, were admixed with the sensitized thymus cells. Both the mixtures and their components alone were plated directly on labeled test monolayers at doses of 2.5 × 10[6] cells/plate for the determination of chromium release.
[b] T° = sensitized rat thymus cells; S^H = heterochimeric spleen cells; S^B = B6D2F1 mouse spleen cells.

TABLE 7. The lytic activity produced by combinations of sensitized thymus and heterochimera spleen cells admixed in different ratios[a]

Cells comprising culture[b] (ratio)	Lysis ± SD	Lytic increase (%)	Significance
T°	12.8 ± 0.4		
S^H	1.2 ± 0.9		
T° + S^H			
10:1	16.6 ± 1.9	38	$P < 0.001$
3:1	21.0 ± 0.2	51	$P < 0.001$
1:1	23.5 ± 1.4	56	$P < 0.001$
1:3	18.6 ± 0.4	45	$P < 0.001$
1:5	18.6 ± 0.8	45	$P < 0.001$
1:10	17.6	41	
S^B	2.6 ± 0.4		
T° + S^B			
3:1	11.3 ± 2.1	0.0	—

[a] The same components were used in these experiments as were described in [a], Table 6. The cells were incubated for 2 days before being transferred to ^{51}Cr-labeled test monolayers for the determination of chromium release.

[b] T° = thymus cells sensitized for 9 days; S^H = heterochimeric spleen cells; S^B = B6D2F1 mouse spleen cells.

strated synergistic cooperation between sensitized rat thymus cells and cells originating from rat bone marrow.

Further experiments tested the quantitative relationship of the interaction. Sensitized thymus and heterochimeric spleen cells were mixed at various ratios ranging from 10:1 to 1:10, so that the final quantity of cells in each culture was 8×10^6 cells. Thus, increasing amounts of chimera spleen cells were added to diminishing amounts of thymus cells. Untreated B6D2F1 spleen cells also were included. An intermediate culture was incubated for 2 days, in this experiment, before we plated the cells on the labeled test monolayers for determination of chromium release.

As seen in Table 7, a dose response relationship was manifested. This relationship resembles that observed in our previous experiments with rat thymus cells admixed with rat spleen cells (22). If we compare the data of Tables 6 and 7, we might consider the optimal ratio to range between 1:1 and 1:3 of sensitized thymus versus heterochimeric spleen cells.

In four experiments (in addition to those described above) using combinations of sensitized thymus and chimeric spleen cells, we observed increased chromium release compared to that of thymus or spleen cells alone. These experiments were made with chimeric spleen cells, 15–62 days after bone marrow inoculation. In the case of only one animal at 12 days after bone marrow inoculation did we fail to demonstrate this effect. In this system, which involved only a short cultivation time, no macrophages were observed.

Is the rat bone marrow graft capable of recognizing host antigens? Having demonstrated cooperation between sensitized rat thymus cells and those deriving from rat bone marrow, we examined whether the rat bone marrow cells or their derivatives could recognize the antigenic environment of their recipients. We tested whether spleen cells of the B6D2F1 chimeras, originating from the rat bone marrow inoculum, were capable of interacting in a reaction directed against histocompatibility antigens present in the B6D2F1 hosts. The following experiment was carried out. A pool of Lewis rat thymus cells was divided into three aliquots. The cells were explanted on C3H (*H-2^k*), C57BL/6 (*H-2^b*), and DBA/2 (*H-2^d*) mouse fibroblast monolayers. After 11 days of culture, the thymus cells were collected. Each batch of sensitized thymus cells was divided and mixed separately, with either heterochimeric (B6D2F1) spleen cells or Lewis rat spleen cells in a thymus:spleen ratio of 3:1. The mixtures and their respective controls (thymus cells, heterochimera, or rat spleen cells alone) were transferred for 2 days to intermediate mouse monolayers, antigenically identical with those on which the given thymus cells were sensitized. The cells were subsequently collected, and the cytolysis was measured on labeled C3H, C57BL/6, or DBA/2 monolayers, respectively.

The rat and the chimeric spleen cells by themselves did not cause significant chromium release. Combinations containing rat spleen cells reacted against all three mouse fibroblast monolayers. On the other hand, the combinations containing chimeric cells reacted only against the C3H monolayer. The fact that the mixtures containing chimeric cells did not react against C57BL/6 and DBA/2 cells raises the possibility that the rat cells of bone marrow origin did recognize and react with the host's tissue antigens (Fig. 2).

The activity of cultures consisting of "sensitized" chimeric cells and fresh thymus cells. In

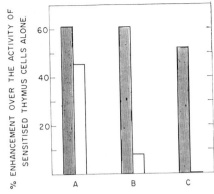

FIGURE 2. Synergistic effect of thymus and heterochimera spleen cell combinations sensitized against host and third party antigens. Rat thymus cells were sensitized on C3H (A), C57BL/6 (B), or DBA/2 (C) mouse embryo fibroblasts (9 days). The thymus cells were mixed in a ratio of 3:1, with either rat or chimera spleen cells (B6D2F1). The combinations were incubated on the respective intermediate monolayers (sensitization against C3H, incubation on C3H, etc.) for 2 days, after which the lysis was determined on the respective labeled fibroblast cultures. The percentage of enhancement measured in the combinate cultures over the chromium release revealed by the respective thymus cultures is plotted. Black columns: sensitized thymus + rat spleen cells; open columns: sensitized thymus + chimera spleen cells.

our previous study (22), we demonstrated that sensitized rat thymus cells combined with nonsensitized rat spleen cells produce a lytic effect enhanced both in time of appearance and extent of lysis. The heterochimeric spleen cells can be regarded as rat bone marrow-derived cells. These cells offered an opportunity to test whether their presensitization would also lead to an enhanced synergistic effect. In the experiment demonstrated in Table 8, in vitro preincubated heterochimeric cells were combined with rat thymus cells. This combination did not elicit any lytic effect.

DISCUSSION

Our knowledge of the lymphoid cell interaction in immune reactions is based mainly on experiments in which thymectomized, irradiated mice are reconstituted to immunological competence by bone marrow and thymus cells (2, 9,

11, 13, 17, 18, 24, 37, 39). To combine bone marrow and thymus cells in our in vitro graft reaction system in an attempt to study a similar interaction seemed to us a logical approach. Yet, combining rat bone marrow cells with either sensitized or nonsensitized thymus cells in tissue culture resulted in no synergistic effect.

There are only a few cases reported in which lymphoid interaction was demonstrated in an in vitro system. Globerson et al. found interaction in a spleen organ culture between bone marrow and thymus cells, resulting in either specific splenomegaly (17) or an antisheep red blood cell response (18). Recently, Doria et al. (13) described synergistic interaction between mouse thymus and spleen cells in an in vitro antisheep red blood cell system, but they showed (much as we did) that bone marrow cells as such do not give a synergistic effect in vitro with thymus cells.

If we compare the studies on lymphoid cooperation in vivo (2, 7, 24, 37, 39) and in vitro (13, 17, 18, 22), it seems evident that synergistic effects between thymus and bone marrow were obtained only when the cells were administered in vivo or were incubated in an organ culture, but were not obtained in the absence of organized tissue structure. This could imply that the so called nonthymus-processed or bone marrow-derived cells in the periphery are not identical with those residing in the bone marrow of small rodents. On the other hand, it appears that lymphoid cells in general derive from a common source—the hematopoietic organs (11, 14, 15, 27, 42, 43). Therefore, it was logical to suppose that the partners of thymus cells in lymphoid cooper-

TABLE 8. The activity of cultures consisting of preincubated (9 days) heterochimera spleen cells and fresh rat thymus cells[a]

Cells comprising culture[b] (ratio)	Lysis ± SD
T	2.3 ± 0.78
S[H]	0.0
T + S[H] (3:1)	0.4 ± 0.47

[a] A spleen cell suspension from heterochimeras was cultured on C3H monolayers for 9 days. The cells then were collected and mixed with rat thymus cells. The mixture and its components alone were transferred directly to labeled test monolayers for the determination of chromium release.

[b] T = thymus cells; S[H] = heterochimeric spleen cells; T + S[H] = thymus + spleen cells.

ation might be differentiation products, independent of the thymus, deriving from the bone marrow. This supposition was tested in the heterochimera experiments. Rat bone marrow cells were injected into thymectomized, lethally irradiated mice. The thymectomy was performed with the aim of ensuring the absence of any possible nonspecies-specific humoral function of the thymus (40). Spleen or lymph node cells of these heterochimeras gave a marked synergistic effect if combined with sensitized rat thymus cells. The reaction was attributable to the presence of rat cells since rat cells were demonstrated in the peripheral blood of the donors and since no lytic effect could be observed when normal mouse cells were combined with the rat thymus cells. The synergistic reaction had a characteristic dose dependence resembling that observed previously in combinations of rat spleen and thymus cells (22).

Our concept is thus supported by two facts. (1) Thymus and bone marrow cells when cultured together in the presence of foreign cells do not produce an immune reaction. (2) When thymus cells are combined with cells of bone marrow origin which have repopulated the spleen of thymectomized recipients, the combination produces a synergistic effect. Hence, the nonthymus-processed cells deriving from the bone marrow need a further step of differentiation (which might take place in some organized lymphoid structure) before achieving a state of competence to interact with thymus-processed cells. This assumption leads to a symmetric concept of lymphoid differentiation. Lymphoid cells derive from the hematopoietic organs but have to differentiate further in either the thymus or the peripheral lymphoid organs in order to become immunologically potent cells.

The lymphoid interaction as a cellular basis for certain thymus-dependent humoral immune reactions is fairly well established (8, 10, 25, 39). In the case of cell-mediated immune reactions, there are still some contradictions in the literature. Argyris (2) showed recently that homograft reactions in thymectomized, irradiated mice can be restored by thymus and bone marrow cells. Stuttman and Good (38) and Tyan (41) did not find lymphoid cooperation, upon injection of bone marrow cells, in the GVH reaction produced by thymus cells. These negative results were confirmed by Davis et al. (12). On the other hand, Globerson and Auerbach (17) in

1967 demonstrated bone marrow-thymus synergism in an in vitro GVH system. More recently, Cantor and Asofsky (7) showed that synergistic interaction takes place in the classical GVH reaction too, if carefully selected ratios of the cells are administered. Very similar findings were reported by Barchilon and Gershon (4). Our results are in agreement with those of Argyris, Globerson and Auerbach, Cantor and Asofsky, and Barchilon and Gershon. The discordance in literature on thymus-bone marrow synergism in cellular immune reactions and the apparent lack of cooperation in the antiflagellin system (3)—a "non-thymus-dependent" antibody reaction—may suggest either that there are immune reactions based on only one type of lymphocyte or, more likely in these cases, that cooperation might occur between two cell populations present in widely differing ratios.

One of the main questions with regard to lymphoid cooperation is whether both cells are capable of recognizing the antigen. That thymus cells can recognize the antigen is generally accepted (11, 20, 25, 26, 37, 39). We have demonstrated that, in our in vitro heterograft system too, the thymus cells are antigen-reactive; they can initiate the reaction and determine its specificity (22). In the present study, we demonstrated that the nonthymus-processed cells also might be antigen-sensitive. This possibility is deduced from the experiment indicating that nonthymus-processed derivatives of bone marrow cells did not effect lysis of fibroblast genetically identical with their chimeric host (C57BL/6 and DBA/2) but did interact in reactions directed against mouse cells of different genetic origin (C3H). This specific unresponsiveness can be explained in several ways.

Van Bekkum (44) demonstrated that long lived bone marrow chimeras possess a donor-type lymphoid system which is specifically tolerant of the host but is reactive to third party antigens. Lengerova and Polackova (21) showed that bone marrow-derived cell populations became tolerant if exposed to cellular antigens in sufficient quantity. Specific adoptive tolerance of bone marrow cells toward allografts or sheep red blood cells was described by Argyris (1) and by Playfair (34), respectively. In contrast to this, several authors could not produce adoptive tolerance by bone marrow cells (20, 26, 39). Generally, it is believed that, if bone marrow (nonthymus-processed) cells can be made tolerant, the tolerance is not long lasting. Apart from tolerance, there is an alternative explanation for the specific unresponsiveness toward host antigens in our chimera cells. In the spleens and, therefore, in the cell suspensions of the heterochimera, a certain amount of host cells were present. The presence of the host cellular antigens might have led to the "neutralization" of antihost potencies in the grafted rat bone marrow cells and, so, only the potencies not complemented by the host antigens might have remained operational. Be the mechanism as it may, the observed specific unresponsiveness clearly suggests that the bone marrow lymphocytes or their derivatives are capable of recognizing antigens.

This finding suggests that in the cooperation system both cells can recognize the antigen. This assumption is strongly supported by the interpretation of the hapten-carrier effect. In the case of immune responses to hapten-carrier conjugates, it has been indicated that the reaction might be operated by two cells, one recognizing the carrier, the other the hapten (29, 36). Very recently, an experimental analogy was presented connecting the hapten-carrier effect and the thymus-bone marrow synergism (35), thus accentuating the view of dual antigen sensitivity in lymphoid cooperation.

Dealing with a cooperation system where both cells are antigen-sensitive, it becomes necessary to know which of the cells initiates the reaction. In a previous work, we showed that sensitized thymus cells, but not sensitized spleen cells, can initiate a synergistic reaction (22). This demonstration was repeated in the present study in which synergistic lytic effect was demonstrated between sensitized thymus cells and nonsensitized bone marrow-derived chimera cells. To test the reverse possibility, we combined sensitized (cultured for 8 days) heterochimeric spleen cells with fresh thymus cells. This combination did not give any immunological activity. Whether this lack of activity demonstrates the exclusive initiating role of thymus cells or whether it was attributable to the appearance of macrophages in the rat bone marrow-derived cultures, which might be the representation of lymphocyte-macrophage transformation (6), is yet not known. The possibility that the initiator cell in immune reactions is exclusively the thymus cell is strongly supported by Shearer and Cudkowicz (37), who found in a double transfer system for sheep hemolysin production that thymus and

not bone marrow contains antigen-reactive cells capable of initiating the response.

In humoral immune reactions, a clear-cut division of labor exists between the interacting lymphocytes regarding the initial antigen reactivity and the final effector function—the antibody production (10, 25, 39). This, however, need not be true for cell-mediated immune reactions. We performed preliminary experiments aimed at identifying the killer cell in graft reactions. These experiments (23) indicated that, in mice, the in vitro specific cytolytic effect exerted by sensitized lymphocytes is dependent on the presence of anti-θ-sensitive cells. This finding implies that in this transplantation reaction either the effector cell is the thymus-processed cell or, if the lysis is mediated by more than one cell type, that the presence of thymus-derived cells is obligatory for the effector function.

Acknowledgments. We wish to express our thanks to Drs. Amiela Globerson and Gideon Berke for stimulating discussions. The idea of using heterochimeras was suggested by Dr. Berke. For the thymectomies, we thank Professor N. Haran-Ghera and Mr. S. Yehezkel. The excellent technical assistance of Mrs. Varda Segal is also gratefully acknowledged.

REFERENCES

1. Argyris, B. F. 1968. J. Immunol. 100: 1255.
2. Argyris, B. F. 1969. Transplantation 8: 538.
3. Armstrong, W. D.; Diener, E.; Shellam, G. R. 1969. J. Exp. Med. 129: 393.
4. Barchilon, J.; Gershon, R. K. 1970. Nature 227: 71.
5. Berke, G.; Ax, W.; Ginsburg, H.; Feldman, M. 1969. Immunology 16: 643.
6. Boak, J. L.; Christie, G. H.; Ford, W. L.; Howard, J. G. 1968. Proc. Roy. Soc. B 169: 307.
7. Cantor, H.; Asofsky, R. 1970. J. Exp. Med. 131: 235.
8. Claman, H. N.; Chaperon, E. A. 1969. Transpl. Rev. 1: 92.
9. Claman, H. N.; Chaperon, E. A.; Triplett, R. F. 1966. Proc. Soc. Exp. Biol. Med. 122: 1167.
10. Davies, A. J. S. 1969. Transpl. Rev. 1: 43.
11. Davies, A. J. S.; Leuchars, E.; Wallis, V.; Marchant, R.; Elliot, E. V. 1967. Transplantation 5: 222.
12. Davis, W. E., Jr.; Cole, L. J.; Schaffer, V. T. 1970. Proc. Soc. Exp. Biol. Med. 133: 144.
13. Doria, G.; Martinozzi, M.; Agarossi, G.; diPietro, S. 1970. Experientia 26: 410.

14. Feldman, M.; Globerson, A. 1964. Ann. N.Y. Acad. Sci. *12*: 182.
15. Ford, C. E.; Hamerton, J. L.; Barnes, D. W.; Loutit, J. F. 1956. Nature *177*: 452.
16. Ginsburg, H.; Sachs, L. 1965. J. Cell. Comp. Physiol. *66*: 199.
17. Globerson, A.; Auerbach, R. E. 1967. J. Exp. Med. *126*: 223.
18. Globerson, A.; Feldman, M. *Proceedings of the conference on mononuclear phagocytes, Leiden, 1969.* Blackwell Scientific Publications, Oxford (in press).
19. Gömöri, L. 1951. J. Lab. Clin. Med. *37*: 526.
20. Isakovic, K.; Smith, S. B.; Waksman, B. H. 1965. J. Exp. Med. *123*: 1103.
21. Lengerova, A.; Polackova, M. 1963. Folia Biol. (Praha) *9*: 189.
22. Lonai, P.; Feldman, M. 1970. Transplantation *10*: 372.
23. Lonai, P.; Clark, W. R.; Feldman, M. Nature (in press).
24. Miller, J. F. A. P.; Mitchell, G. F. 1968. J. Exp. Med. *128*: 801.
25. Miller, J. F. A. P.; Mitchell, G. F. 1969. Transpl. Rev. *1*: 3.
26. Miller, J. F. A. P.; Mitchell, G. F. 1970. J. Exp. Med. *131*: 675.
27. Miller, J. F. A. P.; Doak, S. M. A.; Cross, A. M. 1963. Proc. Soc. Exp. Biol. Med. *112*: 785.
28. Mitchell, G. F.; Miller, J. F. A. P. 1968. J. Exp. Med. *128*: 821.
29. Mitchison, N. A. 1969. p. 149. *In* W. Braun
and M. Landey (eds.). *Immunological tolerance.* Academic Press, New York.
30. Nossal, G. J. V.; Cunningham, A.; Mitchell, G. F.; Miller, J. F. A. P. 1968. J. Exp. Med. *128*: 839.
31. Parkhouse, R. M. E.; Dutton, R. W. 1966. J. Immunol. *87*: 663.
32. Parrott, D. M. V.; deSousa, M. A. B.; East, J. 1965. J. Exp. Med. *123*: 191.
33. Perkins, E. H.; Makinodan, T. 1965. J. Immunol. *95*: 765.
34. Playfair, J. H. L. 1969. Nature *222*: 883.
35. Raff, M. C. 1970. Nature *226*: 1257.
36. Rajewsky, K.; Schirrmacher, V.; Nase, S.; Jerne, N. K. 1969. J. Exp. Med. *130*: 848.
37. Shearer, G. M.; Cudkowicz, G. 1969. J. Exp. Med. *130*: 1243.
38. Stuttman, O.; Good, R. A. 1969. Proc. Soc. Exp. Biol. Med. *130*: 848.
39. Taylor, R. B. 1969. Transpl. Rev. *1*: 114.
40. Trainin, N.; Linker-Israeli, M. 1967. Cancer Res. *27*: 309.
41. Tyan, M. L. 1969. Proc. Soc. Exp. Biol. Med. *132*: 1182.
42. Tyan, M. L.; Cole, L. J. 1965. Nature *208*: 1223.
43. Tyan, M. L.; Herzenberg, L. A. 1968. Proc. Soc. Exp. Biol. Med. *128*: 952.
44. van Bekkum, D. W. 1963. Transplantation *1*: 39.

Received 9 October 1970.
Accepted 24 November 1970.

G. DENNERT
E. LENNOX

Cell Interactions in Humoral and Cell-mediated Immunity

THYMUS-derived lymphocytes (T-cells) play an important role in the initiation of both humoral and cell-mediated immunity[1-3]. We have investigated whether the helper function and the cell-mediated killer function of lymphocyte populations are performed by the same cells, by assaying thymus-derived lymphocytes both for their capacity to cooperate *in vitro* with bone marrow-derived lymphocytes (B-cells) in the induction of plaque-forming cells, and for their capacity to cause *in vitro* complement independent lysis of target cells.

A suitable cell for these two assays is the chicken red blood cell (CRBC) which can be used as immunogen in Mishell–Dutton cultures[4] assaying for plaque-forming response[5], and as immunogen in mice assaying *in vitro* for cell-mediated cytotoxicity using it as target cell[6]. Spleen cells of animals sensitized to CRBC show easily measurable destruction of ^{51}Cr-labelled CRBC (Fig. 1). We show this destruction to be complement-independent by failure to detect complement in the supernatant of the incubation mixture (G. D. and E. L., unpublished results), and by showing that lysis cannot be abolished by addition of 50 µg/ml. carrageenan (ref. 7, and G. D. and E. L., unpublished results).

109

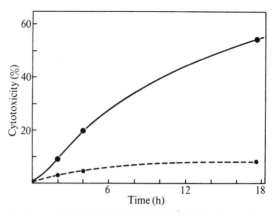

Fig. 1 Target cell lysis by CRBC immunized spleen cells as a function of time. $B_6D_2F_1$ mice (C57Bl $6♀ \times$ DBA♂) were injected with 0.2 ml. of 10% CRBC suspension on day -14. Spleens were minced with forceps in Hanks BSS[4], washed three times in BSS, counted and suspended in Eagle's minimum essential medium for suspension cultures without L-glutamine (Microbiological Assoc., Bethesda, Maryland). This medium was supplemented with 5% foetal calf serum ('Gibco'; immuno-precipitin tested, heated for 1 h at 56° C); glutamine (final concentration 2 mM); penicillin (100 i.u. ml.$^{-1}$) and strepto-mycin (100 μg ml.$^{-1}$). CRBC were taken from Black Jersey Giant roosters by venous puncture into a syringe containing heparin (100 i.u. ml.$^{-1}$). Labelling with ^{51}Cr was done as described by Perlmann[6]. To a total volume of 2 ml. was added 2×10^7 lymphocytes, 2×10^5 ^{51}Cr CRBC and 100 μg carragee-nan. Aliquots of 0.25 ml. were put into 10×100 mm 'Falcon' plastic tubes (No. 2057) flushed with 5% CO_2 in air, stoppered and incubated at 37° C on a rocking platform ('Bellco Glass', Vineland, New York). At specified periods, 1.75 ml. of BSS was added to each tube followed by vortexing and centrifuga-tion (10 min at 2,000 r.p.m.). Aliquots of supernatant (1 ml.) were counted in an automatic well-type gamma radiation counter. Results are expressed as % maximal ^{51}Cr release of samples obtained by adding 1.75 ml. distilled water instead of BSS. The values given are means of duplicates, which usually deviate less than 5% from each other. ●—●, Immune spleen cells; ● - - - ●, normal spleen cells.

Spleen cells, containing thymus-derived lymphocytes sensi-tized with CRBC by injection of thymus cells with antigen into lethally-irradiated recipients, efficiently stimulate the anti-CRBC response of unprimed spleen cell cultures (Table 1*a*). The same batch of sensitized T-cells, however, either alone or admixed with normal spleen cells, is incapable of lysing target cells (Table 1*b*). This clearly shows that the helper cell is not the killer cell itself, nor can it immediately make cytotoxic another cell present in a normal spleen. This suggested that antibody may play an important role in the initiation of the

Table 1

a, Cooperation of CRBC Sensitized T-cells

CRBC sensitized T-cells per culture	Antigen per culture	P.f.c./10^6 recovered cells
—	—	12
—	10^6 CRBC	148
5×10^6	—	44
5×10^6	10^6 CRBC	3,780

10^8 $B_6D_2F_1$ (C57Bl $6 \times$ DBA) thymus cells were injected with 0.2 ml. of a 10% CRBC suspension into lethally-irradiated (1,000 r) $B_6D_2F_1$ mice. On day 8, spleen cells were harvested and mixed with 10^7 unprimed spleen cells and antigen in Mishell–Dutton cultures[4]. The assay of the number of plaque-forming cells was done on day 5.

b, Inability of Cooperating Cells to Lyse Target Cells

Spleen cells per ml.	CRBC sensitized T-cells per ml.	% Cytotoxicity at time 2 h	5 h	17 h
10^7 normal cells	—	1	1	2
10^7 normal cells	5×10^5	1	1	3
7.5×10^6 normal cells	2.5×10^6	2	2	3
5×10^6 normal cells	5×10^6	1	1	2
—	1×10^7	2	1	2

T-cells were sensitized as described in *a* and mixed with various amounts of normal spleen cells and 10^5 labelled CRBC. The assay was as described in Fig. 1.

cytotoxicity reaction. We then showed that mouse anti-CRBC antibody induces spleens containing sensitized T-cells to become cytotoxic (Table 2, exp. 1), but it is obviously not the T-cells in this cell population which become cytotoxic because the effector cell can also be found in spleens of lethally-irradiated animals into which no antigen or thymocytes were injected (Table 2, exp. 2). Further evidence against the T-cell being the killer cell is that antibody renders spleens of thymus-deprived animals (Table 2, exp. 4) even more cytotoxic than it does normal spleens (Table 2, exp. 3) (G. D. and E. L., in preparation). These findings suggest that the killer cell is the radiation-resistant macrophage rendered cytotoxic by the anti-target cell antibody. Using the property of macrophages to adhere to glass and plastic, Shortman *et al.*[8] have devised a very effective column method of depleting spleen cells of macrophages. We put normal spleen cells through such glass bead columns and found that the killer cell is not in the cell fraction that runs through the column but is in the sticking cell population (Table 3, exp. 1) which can be subsequently eluted from the glass beads. The same is true for immune spleens, but in this case the sticking cells, without added anti-

111

Table 2 Induction of Cell-mediated Cytotoxicity by Anti-Target Cell Antibody

Lymphocytes per ml.	Volume and dilution of anti-CRBC serum per ml.	% Cytotoxicity at time		
		3 h	6 h	19 h
Exp. 1				
10^7 t.c.$_{CRBC}$	—	2	2	2
10^7 t.c.$_{CRBC}$	50 μl. 10^{-2}	12	22	35
Exp. 2		2.25 h	6 h	15.5 h
7×10^6 s.c. irr	—	3	3	4
7×10^6 s.c. irr	50 μl. 10^{-2}	12	19	48
Exp. 3		2 h	4.5 h	18 h
10^7 s.c.	—	2	2	7
10^7 s.c.	50 μl. 10^{-2}	26	35	63
Exp. 4				
10^7 b.c.	—	1	3	5
10^7 b.c.	50 μl. 10^{-2}	33	41	86

Various lymphoid populations were assayed for their cytotoxicity with and without added anti-target cell antibody. Anti-CRBC serum was raised by injecting 0.2 ml. of a 10% CRBC suspension bi-weekly into $B_6D_2F_1$ mice for three weeks. The serum had an anti-CRBC agglutination titre of 2^7. t.c.$_{CRBC}$ are CRBC-sensitized spleens containing T-cells (as in Table 1). s.c. irr, spleen cells of animals irradiated with 1,000 r. on day -8; s.c., normal spleen cells; b.c., spleen cells of thymectomized lethally-irradiated and bone marrow-protected animals. The assay was done as described for Fig. 1.

body, lyse target cells. This is as we would expect since both macrophages and antibody-secreting cells are retarded by these columns[8]. Confirming previous work[8], we also showed that the cooperating lymphocytes do not stick to glass beads under the conditions used (G. D. and E. L., unpublished results). We think the effector cell is a macrophage, characterized by radiation resistance, adhesion to glass beads and its presence in spleens of thymus-depleted animals. If this is true, anti-θ serum treatment of immune spleen cells should have no effect on their cytotoxicity. We raised AKR anti-$C_3H\theta$ serum as described by Reif and Allen[9] and showed that it did not lyse plaque-forming cells, but did lyse thymus-derived lymphocytes as assayed by their helper function in Mishell–Dutton cultures (G. D. and E. L., unpublished results). Treatment of immune spleen cells with this anti-θ serum and complement almost completely abolished their cytotoxicity (Table 4, exp. 1); we noted that added anti-target cell antibody restored some of their cytotoxic activity. It was this partial restoration which makes us believe that the inhibiting effect of the anti-θ serum is due to

112

Table 3 Absorption of Effector Cells to Glass Beads

Lymphocytes per ml.	Volume and dilution of anti-CRBC serum per ml.	% Cytotoxicity at time		
Exp. 1		2.5 h	6.5 h	18 h
10^7 normal cells	—	3	4	8
10^7 normal cells	50 µl. 10^{-2}	37	61	85
10^7 non-sticking cells	50 µl. 10^{-2}	8	11	12
5×10^6 sticking cells	—	4	4	6
5×10^6 sticking cells	50 µl. 10^{-2}	14	32	60
Exp. 2		2 h	6 h	16 h
10^7 normal cells	—	2	4	8
10^7 immune cells	—	8	27	53
10^7 immune cells (non sticking)	—	4	8	14
5×10^6 immune cells (sticking)	—	10	25	53

Normal spleen cells (exp. 1) and immune spleen cells (exp. 2) (immunized by injection of 0.2 ml. 10% CRBC on days -64, -41, -23 and -4) were passed through glass bead columns as described by Shortman et al.[8]. Sticking cells eluted as described[8], and non-sticking cells were mixed with 10^5/ml. labelled CRBC and assayed for their cytotoxicity.

Table 4 Effect of Anti-θ Serum on Cell-mediated Cytotoxicity

Lymphocytes per ml.	Volume and dilution of anti-CRBC serum per ml.	% Cytotoxicity at time		
		2 h	6 h	18 h
Exp. 1				
10^7 normal cells	—	2	3	6
10^7 immune cells	—	11	21	42
10^7 immune cells	50 µl. 10^{-2}	16	27	47
10^7 immune cells + anti-θ	—	4	8	13
10^7 immune cells + anti-θ	50 µl. 10^{-2}	10	16	24
Exp. 2				
10^7 normal cells	—	1	2	5
5×10^6 immune cells	—	5	20	33
5×10^6 immune cells	50 µl. 10^{-2}	17	28	45
5×10^6 immune cells + anti-θ	—	4	10	22
5×10^6 immune cells + anti-θ	50 µl. 10^{-2}	16	27	46

Exp. 1—Animals were immunized with 0.2 ml. 10% CRBC on days -69, -46, -28 and -9. Spleen cells, 2×10^7/ml., were incubated with 0.15 ml. anti-θ serum for 30 min at 0° C, washed in BSS and then incubated with 0.15 ml. guinea-pig complement in 1 ml. for 30 min at 37° C. Labelled target cells, 10^5/ml., were mixed with the spleen cells.
Exp. 2—The anti-θ serum was absorbed once with AKR spleen and thymus cells. Animals were immunized on days -64, -41, -23 and -4. Spleen cells, 2×10^7, were incubated with 0.2 ml. anti-θ serum and processed as described for Exp. 1.

113

auto-antibodies[10] with specificities not for the θ antigen. To test this, we absorbed the serum with AKR thymus and spleen once and showed that, while it still inhibited, the cytotoxic activity could be completely restored by anti-CRBC antibody (Table 4, exp. 2). Also, continued absorption of this serum with AKR thymus and spleen decreased its suppressive effect on spleen cytotoxicity to CRBC to 10%, but left the serum with a high titre for killing of thymus-derived helper cells (G. D. and E. L., unpublished results).

We suggest that these experiments show the function of the thymus-derived lymphocyte in cell-mediated immunity to be the triggering of the B-lymphocyte to production of target cell specific antibody. Thus the role of the T-cell is the same whether in cell-mediated immunity or humoral immunity. The effector mechanisms are different. In cell-mediated immunity the antibody opsonizes the target cell, which is then attacked by a macrophage. We titrated anti-target cell serum for its ability to induce cell-mediated cytotoxicity or complement-dependent lysis of target cells and found that concentrations ten times lower than those inducing complement-dependent lysis still induced almost optimal destruction by macrophages (G. D. and E. L., unpublished results). This might explain why in cell-mediated cytotoxicity reactions antibody usually cannot be found. It is interesting that unabsorbed anti-θ serum efficiently inhibits the effector mechanism, that is, the action of the macrophage, and that this inhibition can be reversed by anti-target cell antibody (Table 4). This suggests that enhancing (or blocking) and deblocking antibodies[11-13] present in tumour bearing animals are auto-antibodies and anti-target antibodies, respectively, antagonizing as does the unabsorbed anti-θ serum and augmenting as does anti-CRBC antibody. We show that unabsorbed anti-θ serum obviously must be used with great caution in deciding whether the killer cell is actually a T-cell or not[14,15].

Is the mechanism of cell-mediated lysis of target cells described here special to CRBC as target or is it general? We studied this using Wistar–Furth rat spleen cells sensitized to an F344 tumour. These spleen cells lose their cytotoxic activity if filtered through glass bead columns and the cytotoxic cells can be found in the sticking cell population (G. D. and E. L., unpublished results). It seems therefore that also in this allogeneic tumour system the actual killing is done by antibody and macrophages.

There are several reports which argue against such a generalization. In some of these experiments, anti-θ serum was used to show that the killer cell is a T-cell[14,15], but in the light of our findings, such results have to be interpreted with care. In other experiments, thymus cells have been

114

sensitized to allogeneic antigens by injection into lethally-irradiated animals[16] and spleens from such animals were found to be cytotoxic. This is in contradiction to our experiments with T-cells sensitized to CRBC. But as there is ample evidence that indeed autoantibody forming cells can be found in the thymus[17-20], especially those forming IgG and for which these cell populations have not been tested, these reported experiments do not exclude that the cytotoxicity is mediated by antibody and macrophages.

We are not questioning that the T-cell plays a central role in cell-mediated immunity, but we suggest that the role is essentially the same as in humoral immunity, that is, to cooperate with the B-cell.

We thank Mr R. Austin and Mr D. Wegemer for assistance. Financial support was provided by the National Institutes of Health in a contract from the National Cancer Institute, in a research grant from the National Institute of Allergy and Infectious Diseases, and from funds provided by the Salk Institute by Dr Armand Hammer. G. D. is a fellow of the Jane Coffin Childs Memorial Fund for Medical Research.

1 Miller, J. F. A. P., and Osoba, D., *Physiol. Rev.*, **47**, 437 (1967).
2 Cantor, H., and Asofsky, R., *J. Exp. Med.*, **131**, 235 (1970).
3 Hilgard, H. R., *J. Exp. Med.*, **132**, 317 (1970).
4 Mishell, R. I., and Dutton, R. W., *J. Exp. Med.*, **126**, 423 (1967).
5 Jerne, N. K., Nordin, A. A., and Henry, Claudia, *Cell Bound Antibodies* (edit. by Amos, B., and Koprowski, H.), 109 (Wistar Institute Press, Philadelphia, 1963).
6 Perlmann, P., and Holm, G., *Adv. in Immunol.*, **11**, 117 (1969).
7 Borsos, T., Rapp, H. J., and Crisler, C., *J. Immunol.*, **94**, 662 (1965).
8 Shortman, K. N. W., Jackson, H., Russell, P., Byrt, P., and Diener, E., *J. Cell. Biol.*, **48**, 566 (1971).
9 Reif, A. E., and Allen, J. M. V., *J. Exp. Med.*, **120**, 413 (1964).
10 Raff, M. C., *Transplant. Rev.*, **6**, 52 (1971).
11 Hellström, K. E., and Hellström, I., *Ann. Rev. Microbiol.*, **24**, 373 (1970).
12 Sjögren, H. O., Hellström, I., Bansal, S. C., and Hellström, K. E., *Proc. US Nat. Acad. Sci.*, **68**, 1372 (1971).
13 Bansal, S. C., and Sjögren, H. O., *Nature New Biology*, **233**, 76 (1971).
14 Cerottini, J. C., Nordin, A. A., and Brunner, K. T., *Nature*, **228**, 1308 (1970).
15 Lonai, P., Clark, W. R., and Feldman, M., *Nature*, **229**, 566 (1971).
16 Cerottini, J. C., Nordin, A. A., and Brunner, K. T., *Nature*, **227**, 72 (1970).

[17] Marshall, A. H. E., and White, R. G., *Brit. J. Exp. Pathol.*, **42**, 379 (1961).
[18] Ghossein, H. A., and Tricoche, M., *Nature New Biology*, **234**, 16 (1971).
[19] Wilson, J. D., Warner, N., and Holmes, M. C., *Nature New Biology*, **233**, 80 (1971).
[20] van Furth, R., Schuit, H. R. E., and Hijmans, W., *Immunology*, **11**, 19 (1966).

T and B Cell Cooperation Via a Diffusable Factor

Replacement of T-Cell Function
by a T-Cell Product

A. Schimpl
E. Wecker

We have recently shown that the immune response to sheep red blood cells (SRBC) *in vitro* is T-cell dependent, that is, it can be abrogated by pretreatment of the spleen cells with anti-θ serum and complement[1]. We further reported that reconstitution of the system can be achieved by, among other things, the addition of allogeneic thymocytes[2], while syngeneic thymocytes failed to function as helper cells. Positive allogeneic effects were only obtained if the added thymocytes could recognize the remaining B-cells as foreign. One of the explanations evoked for this allogeneic effect was the participation of

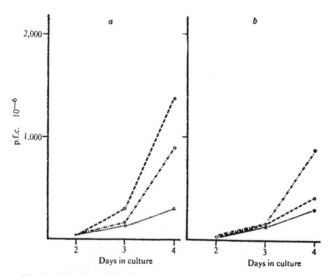

Fig. 1 Kinetics of the immune response to SRBC of partially T-cell deprived spleen cells treated with supernatants from DBA/2 spleen cultures (S-A (DBA)) on day 0 (*a*) and day 2 (*b*). Conditions as described in Table 1. — · — · —, Co; . . . . , anti-θ; - - - -, S-A (DBA).

118

a potentiating factor[2,3]. It seemed possible that on the heavy antigenic stimulation provided by the histocompatibility antigens carried by the B-cells, added T-cells produce a soluble factor which stimulates the immune response of B-cells to unrelated antigens. We now want to report the existence of two distinct factors, one of which can completely replace T-cells.

Experimental procedures were as previously described[1,2,4] or as given in the legends to the figures. Cells from F_1 generations were used to assay for the factors to avoid potentiation by solubilized transplantation antigens which might by themselves stimulate a few remaining T-cells. Essentially, two different factors might account for the allogeneic effect previously reported[2]. (a) A factor might stimulate the proliferation of T-cells and thus lead to functional replenishment of the T-cell pool in a system partially deprived of T-cells (T-cell expanding factor, TEF). This action would be similar to that of several other substances which have already been reported in the literature[6-8]. (b) A factor might actually replace T-cells in their cooperation with B-cells (T-cell replacing factor, TRF). Such a factor might shed some light on the events taking place in normal T-B interaction.

Table 1 Response to SRBC of Partially T-Deprived Spleen Cells in the Presence of Spleen Supernatants

	Recipient cultures	24-h supernatants from	p.f.c. 10^{-6} Day 4	Day 5
DBA	Co	None	3,100	6,800
DBA	Anti-θ serum, complement	None	440	3,900
DBA	Anti-θ serum, complement	A^x	2,980	6,450
F_1*	Co	None	890	2,530
F_1	Anti-θ serum, complement	None	303	1,170
F_1	Anti-θ serum, complement	A^x	1,350	2,100

Supernatants were harvested after 24 h from DBA/2 spleen cultures (S-A (DBA), 1×10^7 cells ml.$^{-1}$; centrifuged for 10 min at $300g$ and 20 min at $2,300g$. Supernatants (0.5 ml.) were added to 0.5 ml. of recipient cultures containing 8×10^6 cells at day 0. The immune response per 10^6 recovered cells was determined at day 4 or 5 by the haemolytic plaque assay[5].
$A^x = DBA/2J$. $F_1* = (DBA/2J) \times (57Bl/6J)F_1$.

In the subsequent experiments the presence of either or both of these factors in supernatants from individual (S-A or S-B) or mixed spleen cultures (S-AB) was studied.

Occasionally we found supernatants from individual spleen cultures which were able to reconstitute a T-deprived system, particularly when anti-θ treatment was incomplete (Table 1). It therefore seemed possible that the activity of these supernatants depends on the presence of a certain number of T-cells which they can expand (TEF). Such a factor may need to

119

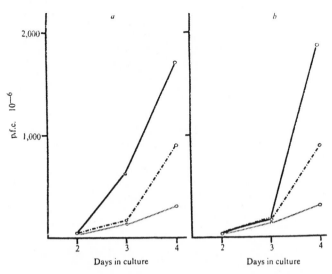

Fig. 2 Kinetics of the immune response to SRBC of a T-cell depleted spleen cell culture treated with supernates from mixtures of allogeneic spleen cells (S-AB) on day 0 (*a*) and day 2 (*b*). Conditions as described in Table 2. — · — · —, Control; . . . ., anti-θ; ——, supernatant AB.

be present from the onset of the cultures so that the T-cell pool can be replenished in time to provide an optimal peak response. Fig. 1 gives the results obtained with S-A added at two different times to a recipient culture which was only partially T-cell deprived. S-A improved the immune response of that culture only if added at day 0, but had little effect if given at day 2.

Table 2 Response to SRBC of T-Cell Depleted Spleen Cultures Treated with Supernatants from Mixtures of Allogeneic Spleen Cells

Spleen cells	24-h supernatants from	p.f.c. 10^{-6} day 4
F_1 GPC*	—	4,450
F_1 GPC	A*	4,120
F_1 GPC	B*	4,500
F_1 GPC	A+B	4,550
F_1 Anti-θ, GPC	—	282
F_1 Anti-θ, GPC	A	510
F_1 Anti-θ, GPC	B	180
F_1 Anti-θ, GPC	A+B	1,980

* A: DBA/2J mice; B: C57Bl/6J mice.
S-AB 5×10^6 spleen cells from DBA/2 and 5×10^6 spleen cells from C57Bl/6J in 1 ml. incubated for 24 h. Supernatants were harvested and centrifuged as described for S-A in Table 1. GPC = Complement.

120

To search for a factor that actually replaces T-cells, recipient cultures were now exhaustively treated with anti-θ serum and complement (Table 2). The residual responsiveness of 5% could not be improved by the addition of S-A at day 0, while S-AB lead to the formation of a significant number of antibody forming cells. This hinted at the presence of a T-cell replacing factor in S-AB. In contrast to TEF, TRF might act even better if added later, particularly if the B-cells had to be provided with appropriately processed antigens before they can respond to it.

Fig. 2 shows that S-AB was indeed increasingly potent in restoring the immune response the later it was added. Thus, either TRF is used up or it is unstable and no longer available when the B-cells are ready.

Rigorously to prove the existence of TRF the supernatants were further tested on spleen cells of the congenitally athymic nude mice[9]. Our previous assumptions were borne out by the experiments (Table 3). S-A and S-B, that is TEF, were totally ineffective, no matter when they were added: no expandable T-cells are present in nudes. S-AB, however, gave excellent responses and again was significantly more effective the later it was added. Fig. 3 shows the dose–response curve of TRF which seems to be present in rather low concentrations.

Table 3 *In vitro* Immune Response of Spleen Cells from Nude Mice (Bomholtgard, Ry, Denmark) to SRBC.

Spleen cells	Supernatants (24 h) from	Time added	p.f.c. day 4
Nude	—	—	175
Nude	A*	Day 0	210
Nude	A	Day 1	250
Nude	A	Day 2	220
Nude	B*	Day 2	157
Nude	A+B	Day 0	625
Nude	A+B	Day 1	1,050
Nude	A+B	Day 2	1,680

---

* A: CBA mice; B: C57Bl/6J mice.

The type of culture from which the supernatants were derived is given in the second column. Supernatants (0.5 ml.) were added at the time indicated to 0.5 ml. of spleen cultures containing 8×10^6 cells.

A factor that substitutes for T-cell function can be supposed to be a T-cell product. S-AB anti-θ (Table 4) was totally ineffective in restoring the response to SRBC. Thus, TRF seems to be the product of T-cells. One might argue that TRF is a B-cell product only formed in the presence of T-cells. In view of the short incubation time needed for its formation and the direction of recognition by thymocytes in the allogeneic reconstitution[2] this seems rather unlikely.

121

Table 4   Evidence for TRF being a T-Cell Product

Spleen cells	24-h Supernatants from	p.f.c. 10^{-6} Day 4	Day 5
F_1 CO	—	480	2,050
F_1 Anti-θ, GPC	—	40	216
F_1 Anti-θ, GPC	A*+B*	254	1,830
F_1 Anti-θ, GPC	A-Anti θ+B-Anti θ	36	174

A* = DBA/2J mice.
B* = C57Bl/6J mice.
24 h supernatants were harvested from allogeneic spleen cell cultures (S-AB) and from anti-θ-treated allogeneic spleen cell cultures (S-AB-anti-θ). 0.5 ml. of recipient cultures (8×10^6) were treated with 0.5 ml. of respective supernatants on day 2.

Several reports deal with substances that can stimulate the growth of T-cells and by doing so restore the helper cell dependent immune response in partially T-deprived *in vitro* and *in vivo* systems[6,7]. In some instances T-cell replacing factors have also been postulated even though no rigorous proof for the total absence of T-cells was given[10-12]. We think that using cell free supernates and the most stringent conditions now available for the absence of T-cells we have proved the existence of a factor with the following characteristics: it is demonstrably produced by T-cells on contact with foreign histocompatibility antigens; if added in sufficient quantities at the right time it can fully substitute for T-cells in the response *in vitro* to unrelated antigens such as SRBC; even under very heavy antigenic stimulation the amount of factor produced is low.

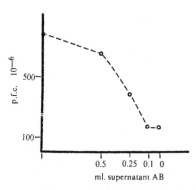

Fig. 3   Dose–response curve of S-AB on nude mice. All volumes were adjusted to 1 ml. by addition of fresh medium. ◯, Nude control.

122

The following tentative model of T-B cooperation would account for all these qualities: T-cells stimulated by their specific antigen release a substance (TRF) which is able to stimulate B-cells to proliferate and/or differentiate. It seems reasonable to assume that it can only do so if a B-cell carries its specific antigen. Because of the low concentration of TRF available, normally only the closest B-cell fulfilling the requirement of carrying antigen on its receptor will respond. This will be the B-cell linked to the T-cell *via* an antigenic bridge. There is therefore no contradiction whatsoever with the original Mitchison–Rajewsky model[13,14] which requires close T-B association brought about by a bivalent antigen. The model suggested here implies an additional, if rare, feature: an antigen carrying B-cell could benefit from a TRF releasing T-cell even if not linked to it or to any other T-cell. Data reported by Katz *et al.*[15], Hartmann[16], Hunter *et al.* (personal communication) and McCullagh[17] could be interpreted in this sense and have led to similar models[11].

Quite different interpretations can also be envisaged especially in view of the fact that this factor, functionally replacing T-cells, has so far only been derived from allogeneic systems.

Further insight regarding its possible physiological role will be gained by using defined T-cell depended antigens, T-cell independent antigens and haptens.

This and the physical and chemical character of the T-cell replacing factor are currently investigated.

We thank Miss L. Hink and Mrs I. Wolf for technical assistance. This work was partially supported by the Deutsche Forschungsgemeinschaft and the Stiftung Volkswagenwerk.

[1] Schimpl, A., and Wecker, E., *Nature*, **226**, 1258 (1970).
[2] Schimpl, A., and Wecker, E., *Europ. J. Immunol.*, **1**, 306 (1971).
[3] Janis, M., and Bach, F. H., *Nature*, **225**, 238 (1970).
[4] Mishell, R. J., and Dutton, R. W., *J. Exp. Med.*, **126**, 423 (1967).
[5] Jerne, N. K., and Nordin, A. A., *Science*, **140**, 405 (1963).
[5] Cone, R. E., and Johnson, A. G., *J. Exp. Med.*, **133**, 665 (1971).
[7] Ishizuka, M., Braun, W., and Matsumoto, T., *J. Immunol.*, **107**, 1027 (1971).
[8] Byrd, W., *Nature New Biology*, **231**, 280 (1971).
[9] Pantelouris, E. M., *Nature*, **217**, 370 (1968).
[10] Ekpaha-Mensah, A., and Kennedy, J. C., *Nature New Biology*, **233**, 174 (1971).
[11] Dutton, R. W., *First Intern. Congr. Immunol.* (Washington DC, 1971).
[12] Diamantstein, T., Wagner, B., L'Age-Stehr, J., Beyse, I., Odenwald, M. V., and Schultz, G., *Europ. J. Immunol.*, **1**, 302 (1971).
[13] Mitchison, N. A., in *Immunological Tolerance*, 113 (Academic Press, New York, 1969).
[14] Mitchison, N. A., Rajewsky, K., and Taylor, R. B., *Prague Symposium on Developmental Aspects of Antibody Formation and Structure* (edit. by Sterzl, J.), 547 (Prague, 1970).
[15] Katz, D. H., Paul, W. E., Goidl, E. A., and Benaceraff, B., *J. Exp. Med.*, **133**, 169 (1971).
[16] Hartmann, K.-U., *J. Exp. Med.*, **133**, 1325 (1971).
[17] McCullagh, P. J., *J. Exp. Med.*, **132**, 916 (1970).

Enhancing Activity of Thymocyte Culture Cell-Free Medium on the *In Vitro* Immune Response of Spleen Cells from Neonatally Thymectomized Mice to Sheep RBC[1]

G. DORIA, G. AGAROSSI AND S. DI PIETRO

Peripheral lymphoid tissues are made up of mixed populations of thymus-dependent and thymus-independent lymphocytes. In the mouse immunized with sheep red blood cells (RBC), specific antibodies are produced by thymus-independent cells upon interaction with thymus-dependent cells and the antigen (1). The nature of this interaction is not understood, but several mechanisms have been proposed (2). Our previous studies (3) have shown that cooperation between thymocytes from normal mice and splenocytes from neonatally thymectomized mice or from thymectomized chimeras (mice thymectomized in adult life, lethally irradiated, and grafted with isogenic bone marrow cells) can occur *in vitro*, because a primary immune response could be induced *in vitro* only if both the thymocytes and thymus-independent cells were cultured with sheep RBC. This *in vitro* system was used to investigate whether receptor sites are involved in the cooperation between thymocytes and splenocytes from neonatally thymectomized mice. It was found that thymocytes lack demonstrable receptor molecules, such as those detected on thymus-independent cells and present in normal serum, for thymocytes were not prevented by rabbit antibodies to mouse whole serum from interacting with antigen and thymus-independent cells (4).

In the present work we investigated whether the cell-free medium of thymocyte cultures can enhance the *in vitro* immune response of thymus-independent cells. It will be shown that such a cell-free medium can replace thymocytes and display a similar helper effect.

The animals used were (C57BL/10 ♀ × DBA/2 ♂)F₁ hybrid and, in a few experiments, DBA/2 inbred mice of either sex. The Mishell

[1] This work was supported by CNEN-Euratom Association Contract. Publication No. 732 of the Euratom Biology Division.

and Dutton technique (5) was used whereby unprimed spleen cells can be stimulated *in vitro* with sheep RBC to give rise to hemolytic plaque-forming cells (PFC). Cells were obtained from 45-day-old normal or neonatally thymectomized (6) mice and prepared as previously described (3). In each experiment of cell cooperation, 1-ml aliquots of separate or mixed isogenic cell suspensions were distributed in four groups of culture dishes as follows: a) 1.5 × 10⁷ nucleated spleen cells from thymectomized mice and 1 × 10⁷ nucleated thymus cells from normal mice; b) 1 × 10⁷ or 2.5 × 10⁷ nucleated thymus cells from normal mice; c) 1.5 × 10⁷ or 2.5 × 10⁷ nucleated spleen cells from thymectomized mice; d) 1.5 × 10⁷ or 2.5 × 10⁷ nucleated spleen cells from normal mice. Antigen (1 × 10⁷ sheep RBC/dish) was added to each group. Control cultures received no antigen. From day 4 on, cells were harvested daily from two dishes of the same group, pooled, washed and assayed for the number of direct PFC by the Jerne and Nordin technique (7). The results reported in Table I confirm and extend our previous finding of *in vitro* cell cooperation (3). The addition of sheep RBC to mixed cell cultures of thymocytes and splenocytes from thymectomized mice elicited an immune response sometimes comparable to that of normal spleen cells. When the two cell populations were stimulated in separate cultures no response of thymocytes was ever observed, while the response of splenocytes from thymectomized mice was in nearly all cases within the range of the PFC background in unstimulated control cultures. Variations of the cultured cell numbers, as outlined in the aforementioned groups b), c) and d), had no appreciable effects on the immune response. This rules out that higher responses in mixed cultures may simply result from greater cell density.

In each experiment of another series, spleen

TABLE I

In vitro immune response to sheep RBC (PFC/culture) resulting from cooperation between thymocytes and spleen cells from neonatally thymectomized mice (Tx)

Exp. No.[a]	Origin of Cells in Culture	Days of Culture				
		4	5	6	7	8
C74	Tx-Spleen + thymus	35	355	1593	6585	3290
	Thymus	0	0	0	0	0
	Tx-spleen	30	154	146	163	35
	Normal spleen	210	405	2170	5250	420
C100	Tx-Spleen + thymus	208	348	665	759	430
	Thymus	0	0	0	0	0
	Tx-spleen	200	365	220	66	310
	Normal spleen	305	1310	2480	3170	595
C78	Tx-Spleen + thymus	679	2190	2280	170	
	Thymus	0	0	0	0	
	Tx-spleen	351	300	390	55	
	Normal spleen	2540	6960	7800	2350	

[a] (C57BL/10 × DBA/2)F_1 mice were used in experiments No. C74 and C100, DBA/2 mice in experiment C78.

cells from thymectomized mice were suspended in fresh medium at the concentration of 3×10^7 nucleated cells/ml. One volume of this suspension was diluted with equal volume of either fresh medium or cell-free medium from cultures of thymocytes maintained *in vitro* for 24 to 40 hr with or without antigen. One ml of each final spleen cell suspension, containing 1.5×10^7 nucleated cells, was cultured with sheep RBC as described above. The cell-free medium was prepared as follows. One-milliliter aliquots of a thymocyte suspension containing 3×10^7 nucleated cells were distributed in two groups of culture dishes, designated T and TS, only one (TS) of which also received antigen (1×10^7 sheep RBC per dish). After 24 to 40 hr the thymocyte fluids from cultures of each group were pooled and centrifuged at $900 \times G$ for 20 min. Each supernatant, designated T or TS cell-free medium, was added to spleen cells as described above. Absence of cells from the designated cell-free medium was inferred from microscopic inspection and from the results of preliminary experiments in which the activity of the cell-free medium was not decreased after a subsequent centrifugation at $12,000 \times G$ for 30 min.

The results from all experiments of this series are reported in Table II. The data are variable but consistently demonstrate that the immunologic capacity of spleen cells from thymectomized mice can be restored *in vitro* by addition of a cell-free medium in which thymocytes have been cultured with or without antigen for 24 to 40 hr. In control cultures, however, T or TS cell-free medium had no appreciable effect on the *in vitro* immune response of normal spleen cells (data not shown). The observed enhancement by cell-free medium was comparable in magnitude to, but displayed a faster onset than, the one produced by thymus cells. As no consistent difference was noticed between the activity of T and TS, the thymocyte culture cell-free medium seems to lack antigen specificity. Thymus specificity of the active factor is suggested by the finding that cell-free medium in which isogenic adult liver cells had been cultured for 24 hr had no enhancing activity on the immune response of spleen cells from thymectomized mice. From day 4 on, no significant difference in numbers of daily recovered nucleated spleen cells was ever observed among cultures that had received fresh medium, T, or TS. This suggests that the thymocyte culture cell-free medium has no evident blastogenic activity on spleen cells in culture. However, a specific mitogenic activity on immunologically competent cells may well

TABLE II

Enhancing activity of thymocyte culture cell-free medium on the in vitro immune response of spleen cells from neonatally thymectomized mice to sheep RBC (PFC/culture)

Exp. No.[a]	Thymocyte Culture Cell-Free Medium	Days of Spleen Cell Culture			
		4	5	6	7
C89A	None[b]	70	25	10	2
	T[c]	3230	8075	3425	190
	TS[d]	3660	4800	1475	305
C89B	None	2	35	5	8
	T	2305	1900	755	135
	TS	2140	1525	390	48
C83	None	100	158	40	25
	T	415	1265	463	30
	TS	554	955	875	53
C100	None	200	365	220	66
	T	783	1370	1280	691
C78	None	351	300	390	55
	TS	1170	1460	670	620

[a] (C57BL/10 × DBA/2)F$_1$ mice were used in all experiments but No. C78 which was performed with DBA/2 mice.

[b] Thymocyte culture cell-free medium replaced by fresh medium.

[c] T, no addition of antigen to thymocyte cultures.

[d] TS, addition of antigen to thymocyte cultures.

explain the enhancement observed and yet be far below detectability by the nucleated cell counting technique, owing to the small percentage of these cells. Another possible action of the cell-free medium might be the activation of

precursor cells into immunologically competent cells, a process which may not require cell division. Moreover, it remains to be determined whether the cell-free medium acts on thymus-independent cells or on a few thymus-dependent cells that escaped neonatal thymectomy (at birth a small number of thymus-derived cells are already present in peripheral organs (8)). The present results raise other interesting problems such as the physicochemical characterization of the active factor released by thymocytes in culture; the identification of the thymus cells (lymphoid or epithelial-reticular cells (9)) producing the active factor; the relationship between this and other thymic factors enhancing the immune response of thymus-independent cells (10, 11). Some of these problems are under investigation.

REFERENCES

1. Miller, J. F. A. P. and Mitchell, G. F., Transplant. Rev., *1:* 3, 1969.
2. Doria, G., Agarossi, G. and Di Pietro, S., Adv. Exp. Med. Biol., *12:* 63, 1971.
3. Doria, G., Martinozzi, M., Agarossi, G. and Di Pietro, S., Experientia, *26:* 410, 1970.
4. Doria, G., Agarossi, G. and Di Pietro, S., J. Immunol., In press.
5. Mishell, R. I. and Dutton, R. W., J. Exp. Med., *126:* 423, 1967.
6. Miller, J. F. A. P., Br. J. Cancer, *14:* 93, 1960.
7. Jerne, N. K. and Nordin, A. A., Science, *140:* 405, 1963.
8. Raff, M. C. and Owen, J. J. T., Eur. J. Immunol., *1:* 27, 1971.
9. Hays, E. F., J. Exp. Med., *129:* 1235, 1969.
10. Osoba, D., Proc. Soc. Exp. Biol. Med., *127:* 418, 1968.
11. Trainin, N., Small, M. and Globerson, A., J. Exp. Med., *130:* 765, 1969.

Procedures for Classifying T and B Cells

T and B Lymphocytes in Mice Studied by Using Antisera Against Surface Antigenic Markers

Martin C. Raff, MD

THE NOTION that there are two distinct classes of peripheral lymphocytes, each with their own functions and properties, has become a paradigm in immunobiology and it is becoming increasingly more difficult to discuss immunologic phenomena without referring to T and B lymphocytes. The two-lymphocyte model is outlined in Text-fig 1. One type of lymphocyte arises as a stem cell in the bone marrow, migrates to the thymus, where it differentiates to a lymphocyte, and then proceeds out to the peripheral lymphoid tissues where it has been called a thymus-dependent, thymus-derived, thymus-processed or *T lymphocyte*. The other type of lymphocyte also arises as a stem cell in the bone marrow and in birds migrates to the bursa fabricius where it differentiates to a lymphocyte and then migrates to the periphery as a bursa-derived lymphocyte. In mammals the pathway is less clear. It may be that the differentiation of stem cell to lymphocyte occurs within the bone marrow, in the peripheral lymphoid tissues or in some intermediate "bursa-equivalent" tissue. What is clear is that this type of peripheral lymphocyte can fully differentiate without having to pass through the thymus, and it has been called a thymus-independent, bone marrow–derived, bursa-equivalent-derived or a *B lymphocyte*.

Unfortunately, T and B lymphocytes are thus far morphologically indistinguishable, and since both types of lymphocytes are always present in any peripheral lymphoid tissue, there is a pressing need for methods that enable one to distinguish and separate T and B cells. Surface antigenic markers are proving to be very valuable in this respect and I will discuss two such markers in mice: the theta (θ) alloantigen,[1] which distinguishes T lymphocytes, and mouse-specific B lymphocyte antigen (MBLA),[2] which distinguishes B lymphocytes. I will also briefly discuss immunoglobulin receptors on T and B lym-

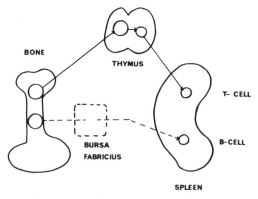

TEXT-FIG 1—The two-lymphocyte model.

phocytes in mice as they too can serve as distinguishing surface antigen markers.

Theta (θ)

In 1963, Reif and Allen, looking for a leukemia-specific antigen, defined an alloantigen that was not only present on a proportion of murine leukemias, but was also present on normal thymus and in nervous tissue.[1] They called the antigen θ and were able to demonstrate that it was determined by a single locus with two alleles: θ AKR found in AKR, RF and a few related substrains, and θ C3H found in all other strains of mice studied. Although the antigen was mainly in brain and thymus, they did find a small amount of the antigen in lymph node and spleen, but it was not clear whether θ in the peripheral lymphoid tissues was carried by all lymphocytes, or whether there was a discrete subpopulation of θ-bearing cells. The finding that only a proportion of peripheral lymphocytes could be killed by anti-θ serum in the presence of guinea pig complement[3] or stained by indirect immunofluorescence with anti-θ,[4] no matter how much antibody was added, suggested that there was indeed a subpopulation of peripheral lymphocytes that was sensitive to anti-θ serum and another that was not. Although it has not been formally established whether the difference between the two populations is quantitative or qualitative, the former is referred to operationally as θ-positive or θ-bearing and the latter as θ-negative or non-θ-bearing. The evidence that the θ-bearing lymphocytes are T cells rests on the observation that mice known on other grounds to be depleted in T cells have very few θ-positive cells in their peripheral lymphoid tissues. This includes mice treated with antilymphocyte serum (ALS),[3,5] neonatally thymectomized mice,[6,7] mice thymectomized as adults, then lethally irradiated and protected with fetal liver or bone marrow cells [7] and congenitally hypothymic ("nude") mice.[7] Also, bone marrow lymphocytes appear to be θ-negative. All the evidence available then suggests that θ-bearing cells are thymus-dependent and that θ can be used as a marker for T lymphocytes. During the past year or so, anti-θ serum has been widely used to study the distribution, differentiation and function of T cells in mice.

The distribution of θ-bearing cells is outlined in Table 1 and is

Table 1—Distribution of θ-Bearing cells in CBA and Balb/c Mice*

Tissue	Percent of θ-positive cells
Thymus	100
Thoracic duct	85
Blood lymphocytes	70
Lymph node	65
Spleen	35
Peritoneal lymphocytes	35
Peyer's patches	30
Bone marrow	0

* Approximate figures obtained by cytotoxicity testing and by indirect immunofluorescence.

generally in accord with estimates of recirculating lymphocytes in mice and rats which are to a large extent thymus-dependent. In CBA and Balb/c mice, there are too few θ-positive cells in the bone marrow to be detected by cytotoxicity or absorption testing. However, chromosome marker studies have established the presence of small numbers of T cells in CBA marrow.[8]

Both θ and the thymus-leukemia (TL) antigen have been used to study the differentiation of T lymphocytes (Text-fig 2). TL is normally present only on thymocytes in TL-positive strains of mice. It is known that in embryonic mice thymocytes develop from stem cells that emigrate from the foetal yolk sac and liver.[9] These cells first arrive in the thymus on day 11 of fetal life and do not express θ or TL on their surface. By day 16, when the first lymphocytes can be seen in the thymus, θ and TL can first be detected, and by day 18 the majority of thymus cells are lymphoid and express θ and TL.[10,11] When a 14-day thymus rudiment, which contains only stem cells and no lymphocytes, θ or TL, is isolated in a cell-tight Millipore chamber on a chick chorioallantoic membrane, the majority of cells are lymphocytes and express both surface antigens 4 days later.[11] It is clear then that the

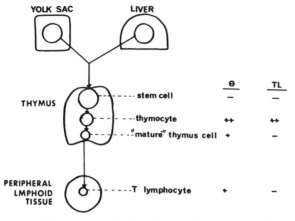

TEXT-FIG 2—The differentiation of T lymphocytes in embryonic mice with associated changes in the surface expression of θ and TL.

acquisition of θ and TL is part of the differentiation of stem cell to thymocyte. The available evidence suggests that the same is true in adult mice, where stem cells migrate to the thymus from the bone marrow.[10,12]

Although thymus cells migrate to the peripheral lymphoid tissues to become T lymphocytes, thymocytes and peripheral T cells are very different. They differ in their immunologic competence and in their surface antigenic characteristics. For example, thymocytes have more θ and less H-2 than do peripheral lymphocytes and only thymocytes normally express TL (Text-fig 2). Thus there must be a further differentiation step between a thymocyte and a thymus-derived T lymphocyte. There is now evidence that at least part of this step occurs within the thymus. This conclusion rests largely on the finding of a small population of cells within the thymus that appears to be indistinguishable from peripheral T cells (Text-fig 2). They are immunologically competent,[13,14] TL-negative,[15,16] have less θ on their surface than the majority of thymocytes,[16] migrate to lymph nodes when injected intravenously[15,17] and are relatively corticosteroid-resistant.[13,14,16] It is unlikely that this small pool of mature T-like cells within the thymus has returned to the thymus as part of the recirculating pool: firstly, because they are not affected by ALS given *in vivo* [18] and secondly, after irradiation, graft-versus-host activity returns first to the thymus and only later to the peripheral lymphoid tissues.[14] All this suggests that the mature population of cells in the thymus have developed from less mature thymic lymphocytes *in situ*. It must be admitted that the possibility of there being two separate pathways of stem-cell dif-

131

ferentiation, one to thymocyte and one to T cell within the thymus, has not been formally excluded. To do this, immature thymocytes would have to be isolated and their differentiation to mature T-like cells demonstrated; thus far, this has not been done.

Anti-θ serum has been used to study the function of T lymphocytes, generally by treating spleen cells *in vitro* with anti-θ and complement and assaying the surviving spleen cells either *in vitro* or after transfer to an irradiated recipient. The immune responses that have been inhibited by such treatment are outlined in Table 2. All cell-mediated

Table 2—Inhibition of Immune Responses by Anti-θ and Complement **In Vitro**

Cell-mediated immune responses
  Graft-versus-host[19]

Cytotoxicity of lymphocytes **in vitro** against allogeneic target cells[20]
  Transfer of contact sensitivity to DNFB[21]
  Transfer of graft rejection[22]
  Mixed lymphocyte reaction[23]
  In vitro response to PHA*[21]

Humoral antibody responses
  Primary and secondary responses to thymus-dependent antigen (**eg**, SRBC, hapten-protein conjugates) **in vivo** and **in vitro**[24-26]

Antigen-binding cells (ABC)
  Antigens where a proportion of ABC are inhibited by anti-θ
    Bovine serum albumin (BSA)[21]
    Maia squinado hemocyanin (MSH)[27]
    Sheep erythrocytes (SRBC)[28-30]
    Chicken erythrocytes[31]
    TIGAL[27]
  Antigens where ABC are not inhibited by anti-θ
    Pneumococcal polysaccharide[32]
    E coli endotoxin[33]
    Polymerized flagellin[34]

* Although not strictly an immune response, the PHA response in mice has been shown to be a T cell response.[8]

immune responses and other responses known to be T cell functions that have been tested have been inhibited. These studies suggest that T cells are important in these responses, but of course they do not exclude the participation of other cell types. The primary and secondary humoral responses to a variety of thymus-dependent antigens (*eg*, SRBC, hapten-protein conjugates) are inhibited by anti-θ, confirming the now generally accepted notion that T cells somehow cooperate with B lymphocytes in humoral immune responses. On the other hand, the 19S humoral response to pneumococcal polysaccharide [32] and polymerized flagellin,[35] both of which are thymus-independent responses, are not

132

affected by anti-θ treatment. This is consistent with the idea that B cells do not require (and possibly do not get) the help of T cells in responding to thymus-independent antigens.

Similar results have been obtained when anti-θ serum has been used to see if any antigen-binding cells (ABC) are θ-bearing. Studies in a number of different laboratories using radiolabeled antigens and autoradiography or rosette technics to detect ABC have shown that with some antigens a proportion of ABC can be inhibited by anti-θ and complement, while with other antigens, none of the ABC appeared to be inhibited (Table 2). Although some of the studies are preliminary and relatively few antigens have been studied, the trend is clear. Those antigens that are thought to be thymus-independent (pneumococcal polysaccharide, E coli endotoxin, polymerized flagellin) seem to be binding only to B cells, while thymus-dependent antigens (BSA, MSH, SRBC) are binding to both B and T cells. With TIGAL, which appears to be somewhat thymus-dependent,[36] a proportion of the lightly labeled cells can be inhibited by anti-θ but not the heavily labeled "hedgehog" cells.[27] It seems then that the degree of thymus dependence or independence of an antigen and its ability to bind to T and B cells may be directly related. The mechanism and nature of the relationship is unclear.

Before finishing the discussion of θ, I should mention some problems that have recently been recognized. It has become apparent that most anti-θ sera raised in conventional inbred mice are contaminated with antibodies directed against surface antigens other than θ. Autoantibodies,[37] anti-Ly antibodies,[38] and as yet uncharacterized antibodies are probably present in most anti-θ sera. However, they are usually not detected in cytotoxicity testing with hamster or guinea pig complement. On the other hand, rabbit complement increases the sensitivity of cytotoxicity testing,[39] and in its presence contaminating antibodies in anti-θ serum will often be detected.[40] There is also recent evidence that the inhibition of early immune rosette-forming cells (RFCs) by anti-θ serum is dependent on cooperation between autoantibody and anti-θ antibody in the serum, and that anti-θ antibodies alone will not inhibit,[40] suggesting that early immune T RFC have less θ on their surface than most T cells. This may explain the findings of Takahashi that anti-θ sera raised in congenic strains of mice (differing only at the θ locus) do not inhibit immune RFCs.[41] Nonetheless, despite these complexities, it seems that most conventionally raised anti-θ sera can be used to selectively kill or inhibit T cells in the presence of guinea pig complement.

Mouse-Specific B Lymphocyte Antigen (MBLA)[2]

MBLA has been defined by an absorbed antilymphocyte serum induced in rabbits by immunizing with lymph node cells from mice that had been thymectomized, irradiated and reconstituted with the fetal liver cells reconstituted. After absorption with mouse liver and RBC and extensive absorption with mouse thymocytes, the antiserum was found to be cytotoxic only for θ-negative (B) lymphocytes.[2] The antigen(s) defined by the antiserum was found to be present in all strains of mice (7/7) tested and absent in rats, hamsters, guinea pigs and rabbits, and thus appears to be mouse specific.

The conclusion that anti-MBLA kills B cells and not T cells rests mainly on the findings that anti-θ and anti-MBLA do not overlap in their cytotoxic activity in sequential ^{51}Cr cytotoxicity testing. Thus anti-MBLA does not kill θ-bearing cells and anti-θ does not kill MBLA-bearing cells. When used together in dye-exclusion cytotoxicity tests, anti-θ and anti-MBLA give additive results (Table 3). As can be seen

Table 3—Dye Exclusion Cytotoxicity Testing with Anti-θ, Anti-MBLA and Anti-θ + Anti-MBLA

Tissue	Cytotoxic Index (%) with		
	Anti-MBLA	Anti-θ	Anti-MBLA + Anti-θ
Thymus	0	100	—
Thoracic duct	14	88	99
Blood lymphocytes	33	71	94
Lymph node	28	61	86
Spleen	56	31	82
Bone marrow	39	0	—

in Table 3, between θ and MBLA most of the lymphocytes in mice can be accounted for. All thymus lymphocytes are θ-positive and MBLA-negative, while in bone marrow most, if not all, of the lymphocytes (which are 30–40% of the nucleated cells in marrow) are MBLA-positive; no θ-positive cells can be detected. In the rest of the lymphoid tissues, the percentages of cells killed by anti-θ and anti-MBLA are complementary and add up to almost 100%. Since myeloma cells and the majority of plaque-forming cells appear to be sensitive to anti-MBLA, it seems likely that antibody-forming cells, as well as resting B lymphocytes, are MBLA-positive. In a preliminary survey of a number of murine lymphomas using dye-exclusion cytotoxicity testing, 4 of 9 were MBLA-positive.[42] Interestingly, 2 of these 4 were also θ-positive.

Studies using anti-MBLA to inhibit anti-SRBC rosette-forming cells (RFCs) from mice immunized with SRBC have shown that anti-MBLA inhibits only those RFCs which are not inhibitable by anti-θ; when used together with anti-θ, virtually 100% of the immune RFCs can be inhibited.[21] On the other hand, when RFCs from unimmunized mice are studied, anti-MBLA and anti-θ overlap significantly in their activity, suggesting that anti-MBLA is inhibiting some background T RFCs as well as B RFCs.

Attempts to inhibit the ability of spleen cells to adoptively transfer a primary or secondary response to SRBCs or a secondary response to dinitrophenylated bovine serum albumin have all been unsuccessful, and in some cases the response has been enhanced. The inability of anti-MBLA to inhibit B cell function although it is able to inhibit B antigen–binding cells is an unsolved paradox. There are a number of possible explanations:

1. B memory cells may be an anti-MBLA-resistant subpopulation of B lymphocytes.

2. The anti-MBLA serum, which is of low titer, probably does not kill all the B cells when large numbers of cells are treated, and the residual B cells may be stimulated by subcytotoxic concentrations of the antiserum.

3. The activity of residual B cells that have not been killed by anti-MBLA may be enhanced by some adjuvant-like action of the serum; perhaps the numerous thymocytes used for absorption release soluble "activator" factors into the serum.

Until this point is settled, anti-MBLA will have very limited usefulness as a marker of B cells.

Various different methods of preparing anti-MBLA sera have been, and still are being tested. Thus far, the method described in this paper has given the best results. However, preliminary studies with an antiserum raised in rats look very promising.

Surface Immunoglobulin (Ig) Receptors

The "receptor hypothesis" suggests that antigen-sensitive cells have antibody-like molecules on their surface that serve as receptors for antigen.[43] Since T and B cells can both carry immunologic memory,[24] become immunologically tolerant[44] and bind antigen specifically,[45] a corollary of the receptor hypothesis must be that both T and B cells have such receptors.

There are two ways of demonstrating Ig receptors on lymphocytes, both making use of anti-Ig antibodies:

Direct methods, which use anti-Ig antibodies to directly demonstrate Ig determinants (not necessarily receptors) on the surface of lymphocytes, such as immunofluorescence, immunoautoradiography, mixed agglutination, immunoferritin electron microscopy and lymphocyte transformation.

Indirect methods, which use anti-Ig antibody to inhibit the immunologic activity of lymphocytes or their binding of antigen.

The available evidence suggests that the direct methods are for the most part demonstrating Ig determinants on B lymphocytes but not on T lymphocytes.[4] For example, when immunofluorescence is used to demonstrate Ig on the surface of mouse lymphocytes, only a proportion of the cells are positive and this proportion, which is characteristic for each tissue, cannot be increased by increasing the amount of anti-Ig.[46] Also, the percentage of positive cells is the same when immunofluorescence or immunoautoradiography is used, and the latter is several orders of magnitude more sensitive than fluorescence.[46] Thus, there is clearly a population of lymphocytes with readily demonstrable Ig on their surface and another population with a great deal less, or no surface Ig. There are three lines of evidence that suggest that the cells with the readily demonstrable surface Ig are mostly, if not entirely, B cells:

1. The distribution of surface Ig-bearing cells is inversely related to that of θ-bearing cells. For example, in thoracic duct, where 80–90% of the cells are θ-positive, only about 5% have demonstrable surface Ig, suggesting that the majority of T cells are not staining.[4]

2. In ALS-treated, or thymectomized, irradiated and marrow-reconstituted mice, the percentage of surface Ig-positive cells is markedly increased.[4]

3. After treatment of lymph node or spleen cells with anti-θ and guinea pig complement, more than 90% of the residual living cells are surface Ig-positive, while after treatment with anti-MBLA less than 1% are positive.[42]

On the other hand, in some laboratories anti-Ig sera have been shown to inhibit the immunologic function of T lymphocytes, and θ-positive rosette-forming cells can be inhibited by some anti-Ig sera (reviewed in Greaves[45]). Thus, it seems likely that both T and B lymphocytes have Ig receptors on their surface, but that B lymphocytes have a good deal more.

Summary

In the mouse, T and B lymphocytes can be distinguished by means of their differing surface antigens. The theta alloantigen can serve as a marker for T lymphocytes and is being widely used to study the differentiation, distribution and function of T lymphocytes. B lymphocytes carry a distinctive heteroantigen, mouse-specific B lymphocyte antigen (MBLA) and also have readily demonstrable Ig determinants on their surface.

References

1. Reif AE, Allen JMV: The AKR thymic antigen and its distribution in leukemias and nervous tissues. J Exp Med 120:413–433, 1964

2. Raff MC, Nase S, Mitchison NA: Mouse-specific bone marrow-derived lymphocyte antigen as a marker for thymus-independent lymphocytes. Nature 230:50–51, 1971

3. Raff MC: Theta isoantigen as a marker of thymus-derived lymphocytes in mice. Nature 224:378–379, 1969

4. *Idem:* Two distinct populations of peripheral lymphocytes in mice distinguishable by immunofluorescence. Immunology 19:637–650, 1970

5. Schlesinger M, Yron I: Antigenic changes in lymph node cells after administration of antiserum to thymus cells. Science 164:1412–1413, 1969

6. Schlesinger M, Yron I: Serologic demonstration of a thymus-dependent population of lymph node cells. J Immun 104:798–804, 1970

7. Raff MC, Wortis HH: Thymus-dependence of θ-bearing cells in the peripheral lymphoid tissues of mice. Immunology 18:931–942, 1970

8. Doenhoff MJ, Davies AJS, Leuchars E, Wallis V: The thymus and circulating lymphocytes in mice. Proc Roy Soc Lond B 176:69–85, 1970

9. Owen JJT, Ritter MA: Tissue interaction in the development of thymus lymphocytes. J Exp Med 129:431–442, 1969

10. Schlesinger M, Hurvitz D: Serological analysis of thymus and spleen grafts. J Exp Med 127:1127–1137, 1968

11. Owen JJT, Raff MC: Studies on the differentiation of thymus-derived lymphocytes. J Exp Med 132:1216–1232, 1970

12. Boyse EA, Old LJ: Some aspects of normal and abnormal cell surface genetics. Ann Rev Genet 3:269, 1969

13. Blomgren H, Anderson B: Evidence for a small pool of immunocompetent cells in the mouse thymus. Exp Cell Res 57:185–192, 1969

14. Anderson B, Blomgren H: Evidence for a small pool of immunocompetent cells in the mouse thymus: its role in the humoral antibody reponse against SRBC, bovine serum albumin, ovalbumin and the NIP determinant. Cell Immun 1:362–371, 1970

15. Raff MC: Heterogeneity of thymocytes: evidence for a subpopulation of mature lymphocytes within mouse thymus. Nature New Biol 229:182–183, 1971

16. Raff MC, Owen JJT: The use of surface alloantigenic markers to study the differentiation of thymus-derived lymphocytes in mice, Morphological and Functional Aspects of Immunity. Edited by K Lindahl-Kiessling, G Alm, MG Hanna Jr. New York, Plenum Press, 1971, pp 11–15

17. Lance EM, Taub RN: Segregation of lymphocyte populations through differential migration. Nature 221:841–843, 1969

18. Raff MC, Nehlsen S: Unpublished observations

19. Cantor H: The effect of anti-θ serum on the graft-versus-host response of spleen and lymph node cells (unpublished observations)

137

20. Cerottini JC, Nordin AA, Brunner KT: Specific *in vitro* cytotoxicity of thymus-derived lymphocytes sensitized to alloantigens. Nature 228:1308–1309, 1970

21. Greaves M: Personal communication

22. Möller E: Personal communication

23. Mosier D, Cantor H: Personal communication

24. Raff MC: Role of thymus-derived lymphocytes in the secondary humoral immune response in mice. Nature 226:1257–1258, 1970

25. Schimple A, Wecker E.: Inhibition of the *in vitro* immune response to SRBC by treatment of spleen cell suspensions with anti-θ serum. Nature 226:1258–1259, 1970

26. Takahashi T, Carswell EA, Thorbecke GJ: Surface antigens of immunocompetent cells. I. Effect of θ and PC1 alloantisera on the ability of spleen cells to transfer immune responses. J Exp Med 132:1181–1190, 1970

27. Roelants G: Personal communication

28. Bach JF, Muller JW, Dardenne M: *In vivo* specific antigen recognition by rosette-forming cells. Nature 225:1251–1252, 1970

29. Schlesinger M: Anti-θ antibodies for detecting thymus-dependent lymphocytes in the immune response of mice to SRBC. Nature 226:1254–1256, 1970

30. Greaves MF, Möller E: Studies on antigen binding cells. I. Origin of reactive cells. Cell Immun 1:372–385, 1970

31. Bach JF: Personal communication

32. Howard J: Personal communication

33. Sjöberg O, Möller G: The presence of antigen-binding cells in tolerant animals. Nature 228:780–781, 1970

34. Ada G, Raff MC: Unpublished observations

35. Diener E, O'Callaghan FB: Immune response *in vitro* to S Adelaide H-antigens, not affected by anti-theta serum (unpublished observations)

36. Wilcox, HNA, Humphrey JH: Personal communication

37. Boyse EA, Bressler E, Ikitani C, Landes M: Cytotoxic γM autoantibody in mouse alloantisera. Transplantation 9:339–341, 1970

38. Boyse EA, Miyazawa M, Aoki T, Old LJ: Ly-A and Ly-B: two systems of lymphocyte isoantigens in the mouse. Proc Roy Soc B 170:175–193, 1968

39. Boyse EA, Hubbard L, Stockert E, Lamm ME: Improved complementation in the cytotoxic test. Transplantation, 10:446–449, 1970

40. Greaves MF, Raff MC: The specificity of anti-θ sera in cytotoxicity and functional tests of T lymphocytes. Nature (in press)

41. Takahashi T: Personal communication

42. Raff MC: Unpublished observations

43. Mitchison NA: Recognition of antigen. Symp Int Soc Cell Biol 7:29–42, 1968

44. Chiller JH, Habicht GS, Weigle WC: Cellular sites of immunologic unresponsiveness. Proc Nat Acad Sci USA 65:551–556, 1970

45. Greaves MF: Biological effects of anti-immunoglobulins: evidence for immunoglobulin receptors on "T" and "B" lymphocytes. Transplant Rev 5:45–75, 1970

46. Raff MC, Sternberg M, Taylor RB: Immunoglobulin determinants on the surface of mouse lymphoid cells. Nature 225:553–554, 1970

Some of the work described was done in collaboration with others, particularly Drs J. J. T. Owen, N. A. Mitchison, H. H. Wortis and Miss S. Nase. I am grateful to Miss P. Chivers for able technical assistance.

T AND B LYMPHOCYTES IN NEW ZEALAND BLACK MICE

AN ANALYSIS OF THE THETA, TL AND MBLA MARKERS

B. H. WAKSMAN, M. C. RAFF and JUNE EAST

SUMMARY

The lymphocytes of thymus, blood, spleen, lymph nodes and bone marrow were studied in NZB mice between the ages of 1 and 14 months, and compared with lymphocytes of A and CBA strain mice of the same age. By standard cytotoxicity techniques, the proportion of cells possessing the θ, TL and MBLA markers was found to be similar in NZB in mice and controls. In 14-month old NZBs, whose spleen was largely replaced by reticulum cell sarcoma, and in younger recipients of the passaged tumour, there was a reduction in the percentage of θ and MBLA-positive cells in the spleen. In a few mice at 4 and 9 months, small numbers of MBLA-positive cells were present in the thymus and there was a corresponding decrease in θ-positive cells. TL-positive cells were not present outside the thymus, and θ-positive cells were not present in the bone marrow in unusual numbers. NZB peripheral lymphocytes appeared to have the same surface concentration of θ as those of A or CBA mice, as judged by anti-θ titration curves. The reticulum cell sarcoma was shown to be θ-negative and MBLA-negative, while an NZB thymoma was θ-positive, TL-positive and MBLA-negative. It was concluded that the peripheral lymphoid organs contain a large population of T lymphocytes of abnormal character.

INTRODUCTION

The NZB mouse and its hybrids with related strains such as NZW have the interesting property of spontaneously forming a variety of autoantibodies (Mellors, 1967; Howie & Helyer, 1968; East, 1970) and thus provide suggestive models for certain 'autoimmune' diseases in man. All NZB animals produce autoantibody directed against erythrocytes, detectable as a positive direct Coombs' reaction from 4–6 months, and develop haemolytic anaemia. They also have a moderate incidence of antinuclear antibodies (ANA) possibly related to the 'immune complex' glomerulonephritis that appears in older animals. Severe kidney disease, rather than autoimmune haemolytic anaemia, characterizes B/W hybrids, all of which have high titres of ANA and die with renal failure before they are a year old. The process of autoantibody formation may be accelerated by reticuloendothelial stimulants such as *C. parvum* (Halpern & Fray, 1969) or 'immunization' with various materials (Lambert & Dixon, 1968; Steinberg, Baron & Talal, 1969). It can be prevented or suppressed with ACTH, antimetabolites or antilymphocyte globulin (Howie & Helyer, 1965; Russell, Hicks & Burnet, 1966; Denman, Denman & Holborow, 1967). Ultimately, generalized reticulum cell neoplasia, probably originating in the spleen, affects the majority of NZB mice from the age of 11 months. A very few ($<7\%$) animals develop lymphocytic leukaemia between 3 and 8 months of age. Both types of malignancy can be transferred successfully by cell passage to young adult NZB recipients (see East, 1970).

Several recent studies have established that, during the first months of life, NZB and B/W mice may show increased responsiveness to such antigens as bovine serum albumin or sheep erythrocytes (Playfair, 1968; Cerottini, Lambert & Dixon, 1969; Morton & Siegel, 1969). Simultaneously with or following the appearance of autoimmune phenomena, there is a progressive loss of immune responsiveness which affects both antibody-formation (Diener, 1966) plus such thymus-dependent lymphocyte functions as graft-*versus*-host competence (Stutman, Yunis & Good, 1968) and the ability to give a mixed lymphocyte reaction in culture (Leventhal & Talal, 1970). There is also a relative insusceptibility to tolerance, which appears within 6 weeks after birth (Weir, McBride & Naysmith, 1968; Staples & Talal, 1969b). Tolerance to such proteins as bovine γ-globulin, if induced in early life, disappears much more rapidly than in other mouse strains (Staples & Talal, 1969a). However, if tolerance is induced with the use of cyclophosphamide, recovery of specific reactivity is slower than in controls (Russell & Denman, 1969; Staples & Talal, 1970).

While several of these findings have proved difficult to reproduce by different investigators (see reviews cited above), they nevertheless suggested an abnormality of lymphocytic function. We were therefore prompted to undertake the following study of lymphocytic subpopulations which can be identified in NZB mice by the use of the specific antigenic markers θ, TL and MBLA (reviewed in Raff, 1971).

MATERIALS AND METHODS

Animals

Inbred NZB mice were from a colony maintained at the Imperial Cancer Research Fund and fully described in earlier publications (see East, 1970). All other mice were obtained from the breeding unit at the National Institute for Medical Research.

Tumours

Reticulum cell sarcomas were studied in two 14-month old NZB mice, in which they had appeared spontaneously, and also in 4-month old NZB recipients injected intraperitoneally 3–4 weeks earlier with the 42nd, 48th and 49th passage of spleen cells from another sarcoma-bearing donor. A thymoma was studied in its 60 and 68th passage, about 3 weeks after intraperitoneal injection in 4-month old syngeneic recipients.

Cell suspensions

All suspensions were prepared in veronal-buffered saline containing $0 \cdot 1\%$ bovine serum albumin (see Raff & Owen, 1971). They were passed once through short columns of glass wool to remove dead cells, debris and macrophages. In a few instances, a second passage was required.

Antisera

Anti-θ C3H sera were prepared and tested as previously described (Raff & Wortis, 1970). Anti-TL, kindly provided by Dr E. A. Boyse, was prepared in congenic A/TL-negative mice against A strain leukemia and detected only TL (specificities 1, 2 and 3). Anti-MBLA serum was prepared by immunizing rabbits with lymph node cells from thymectomized, irradiated CBA mice restored with foetal liver, and absorbing repeatedly with CBA thymus. The properties of this serum are described in (Raff, Nase & Mitchison, 1971). All sera were heat inactivated and used at concentrations well above the minimum required for maximal lysis of cells bearing the corresponding antigen, i.e. anti-θ 1:4, anti-TL 1:4 and anti-MBLA 1:2.

Complement

In earlier experiments, all cytotoxicity tests on thymus made use of hamster complement absorbed with mouse liver and spleen, at a final concentration of 1:21, while peripheral cells were tested with guinea-pig complement absorbed with mouse erythrocytes, at a final dilution of 1:15. In later tests, guinea-pig complement absorbed with agarose was employed for all cell types at a final concentration of 1:12 (Cohen & Schlesinger, 1970).

Cytotoxicity testing

The trypan blue and Cr^{51} release techniques used are described in detail in earlier publications (Raff & Owen, 1971; Raff, 1969). Data are reported as a 'cytotoxic index' calculated from the actual percentages of cells lysed as follows:

$$\text{Cytotoxic index} = \frac{\% \text{ Lysed with antiserum} - \% \text{ lysed with normal mouse serum}}{100\% - \% \text{ Lysed with normal mouse serum}}.$$

Absorption

Fixed volumes of anti-θ serum (0·1 ml, diluted 1:200) were incubated with 0·1 ml of doubling dilutions of test cells, beginning at 500×10^6/ml, at 37°C for 30 min, then spun and the supernatants tested on Cr^{51}-labelled A strain thymus cells by the usual cytotoxicity technique.

Histology

Paraffin sections stained with haematoxylin-eosin and methyl green-pyronine were prepared from the tissues of two series of NZB and control animals of different ages, and additional spleen sections were examined from a 14-month old NZB and some of the reticulum cell sarcoma recipients. Cytocentrifuge smears, prepared from the suspensions used for cytotoxicity testing, were stained with Giemsa and counted in the usual manner.

RESULTS

Cell suspensions

As reported by many investigators, the spleen and lymph nodes were of normal size in

142

young NZB mice, but were consistently larger than the corresponding organs in A or CBA mice at 4, 9 and 12–14 months. The cell yields from NZB were similarly increased (see Zatz, Mellors & Lance, 1971). Smears of cell suspensions showed a high proportion of large and especially medium-sized lymphocytes in both organs as early as 4 months (Table 1). In some 14-month old animals, there was partial or complete replacement of normal splenic tissue by neoplastic reticulum cells.

TABLE 1. Distribution (%) of cell types in NZB spleen and lymph nodes

			Lymphocytes			
Age	Mouse number	Reticulum cells	Large	Medium	Small+ normoblasts	Other
Spleen						
4 months	1	2	5	38	47	6
	2	0	6	27	60	7
	3	1	3	36	53	7
	4*	63	0	7	24	7
	A†	1	3	12	76	8
9–11 months	5	8	11	38	36	6
	6	1	4	21	71	4
	7	0	6	16	75	3
	8	9	7	25	59	2
	9	6	2	51	38	2
	A†	1	3	13	81	3
14 months	10	48	12	12	21	8
	11	24	3	8	58	6
Lymph node						
4 months	1	0	0	26	69	8
	3	1	1	40	57	2
	4*	0	5	35	59	0
	A†	1	2	8	90	0
9–11 months	5	0	7	34	58	1
	6	0	3	26	72	0
	7	0	0	36	58	6
	9	0	4	40	55	1
	A†	2	1	11	84	2
14 months	10	2	4	36	3	6
	11	0	1	18	81	0

* Recipient of reticulum cell sarcoma.
† Comparable counts in animals of the A strain.

Proportion of cells bearing θ, TL, and MBLA

The results of cytotoxicity testing with anti-θ serum are presented graphically in Fig. 1. Essentially normal values were obtained throughout with the following exceptions. In the thymus, at 4 and 9 months, some animals showed a slight diminution in the percentage of θ-positive cells. In two of four animals tested at 14 months, the spleen contained virtually

143

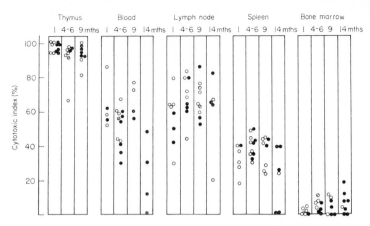

FIG. 1. Cell lysis produced by anti-θ serum in lymphoid populations of NZB, A or CBA mice of various ages. ⊘, Coombs'-positive NZB; ○, Coombs'-negative NZB; ●, control mice. Plotted values represent cytotoxic index

$$= \frac{\% \text{ Lysis by anti-}\theta - \% \text{ lysis by normal mouse serum}}{100\% - \% \text{ Lysis by normal mouse serum}}.$$

no θ-positive cells and the blood as well contained very few cells positive for θ. In these animals, the spleen was almost entirely replaced by malignant reticulum cells. No significant difference was observed between values obtained in Coombs'-positive and Coombs'-negative mice. The bone marrow contained a low percentage of θ-positive cells, but not significantly more than the marrow of control mice.

An absorption study confirmed the presence of θ-positive cells in the bone marrow. Absorption with a sufficiently large number of marrow cells, NZB or A, effectively reduced the lytic activity of a standard amount of anti-θ (Fig. 2).

Tests with anti-TL serum failed to reveal any significant difference in cell distribution from that observed in A or CBA mice (Table 2). TL-positive cells were not found outside the thymus.

In a limited number of cytotoxicity tests with anti-MBLA serum, essentially normal values were obtained for each organ at each age, with two exceptions. In the thymus at 4 and 9 months, several animals (three out of eight) showed positive cells in excess of 10%. These were thought to account for the diminished proportion of θ-positive cells and to correspond to plasma cells seen in moderate numbers in stained sections. In one 14 month spleen replaced by reticulum cell sarcoma, there were no MBLA-positive cells at all.

Concentration of θ on peripheral T-lymphocytes

Titration curves with anti-θ on thymus, spleen, and lymph node cells of five NZB mice, measured by Cr^{51}-release, could not be distinguished from curves obtained with the corresponding cells of A strain mice (Fig. 3a and b). The calculated difference between 50% lysis points for thymus and lymph node or thymus and spleen also were closely comparable in NZB and controls.

144

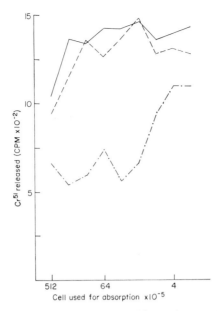

Fig. 2. Absorption of lytic capacity of standard anti-θ serum (at 1:200 dilution) by varying numbers of NZB bone marrow (——), A bone marrow (– – –), and A thymocytes (– · – ·). Cells used for absorption taken from 9-month old donors. Absorbed serum tested on 4 month A thymocytes.

TABLE 2. Percent* of cells lysed by anti-TL

Cells tested	Age of test animals (months)			
	1–2	4–6	9	14
Thymus	91, 87, 85, 59	77, 76, 71	89, 81	—
	69	73	78	
Blood	16, 7, 6	3, 0, 0, 0	—	—
	0, 0	20, 2, 0, 0, 0, 0, 0	0	
Lymph node	8, 7, 0, 0, 0	0	6, 5, 0	15
	4, 0, 0	0	6, 3, 0	0
Spleen	11, 3, 1, 0, 0	6, 2, 2	6, 0, 0	0
	0	6, 0, 0, 0, 0, 0	6, 2	7, 4
Bone marrow	7, 2, 1, 0, 0	15, 10, 5	6, 0, 0	1
	0	11, 7, 3, 0, 0, 0	6, 0, 0	0, 0, 0

* Cytotoxic index. Figures for NZB animals above; A or CBA below.

145

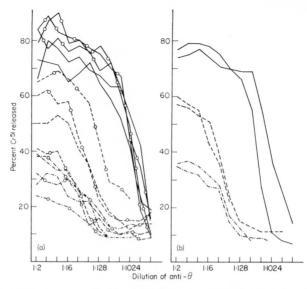

Fig. 3. (a) Titration curves of anti-θ against thymus (——), lymph node (– –), and spleen (– · – ·) cells of NZB mice. Five to six week animals indicated by symbol ○; others were 4 months of age. One animal in each group was male, others female. (b) Similar titration curves against cells of 3 and 4 month female A strain mice.

TABLE 3. Presence of θ and MBLA on tumour cells of NZB mice

Organ tested	Antigen	Cytotoxic index		
		14 month mice with RCS	4 month recipients of RCS	4 month recipient of Th
Thymus	θ	— —	96	—
	MBLA	— —		
Blood	θ	12, 0	60	—
	MBLA	23		
Lymph node	θ	65, 20	56	—
	MBLA	28		
Spleen	θ	0, 0	11, 1	98
	MBLA	1, –	–, 9	0

RCS: Reticulum cell sarcoma; Th: Thymoma.

146

Antigenic character of NZB tumours

Two tumours characteristic of NZB mice were also examined. A reticulum cell sarcoma was clearly negative for both θ and MBLA, while a thymoma was θ positive, TL-positive and MBLA-negative (Table 3). In 14-month old NZB animals, which had spontaneously developed reticulum cell neoplasia, there was a virtually complete disappearance of cells bearing either antigen from the spleen (Fig. 1, Table 3).

DISCUSSION

Our main finding is that, in the thymus and peripheral lymphoid organs of NZB mice, the proportion of lymphoid cells which can be assigned to the categories T (thymus-derived) or B (bone marrow-derived) on the basis of their surface antigenic properties does not differ significantly from the proportions seen in control mice of the A and CBA strains. The characteristic difference in θ concentrations between thymocytes and peripheral T-cells was also found in NZB mice of various ages. The only exceptions were provided by very old animals, in which the progressive replacement of splenic tissue by neoplastic reticulum cells resulted in a depression of both θ- and MBLA-bearing cells, and occasional younger mice, in whom the presence of plasma cell aggregates in the thymic medulla was reflected in the presence of a small proportion of MBLA-positive cells in this organ.

This apparently negative observation assumes significance against the background of recently published data, which suggest that NZB mice may show marked abnormalities in functions usually ascribed to T lymphocytes. Older animals have a deficiency in long-lived lymphocytes (Denman & Denman, 1970) and in cells capable of recirculating to the lymph nodes (Zatz *et al.*, 1971), producing a graft-*versus*-host reaction, or reacting *in vitro* to PHA or an allogeneic stimulus (Stutman *et al.*, 1968; Leventhal & Talal, 1970). Since there is no reduction in the numbers and proportions of lymphoid cells in the spleens and lymph nodes of 9–11-month old NZB mice, one is forced to conclude that most of their θ-positive cells are in fact abnormal, either nonfunctional or preneoplastic cells or possibly functional T-cells engaged in a response to exogenous or autoantigens. In smears of both organs, there was an unusually high proportion of large and especially medium-sized lymphocytes, many basophilic, which could represent such cells.

The preponderance of available evidence is consistent with the hypothesis that the peripheral lymphoid organs of NZB mice contain a large T lymphocyte population engaged in an immune response to autoantigens. First is the fact that their inability to develop tolerance, appearing at about 6 weeks of age (see Weir *et al.*, 1968; Staples & Talal, 1969a and b), is followed by the appearance of autoimmunization, starting as early as 3–4 months in some animals (see Mellors, 1967; East, 1970). Playfair and his colleagues have shown that the NZBs insusceptibility to tolerance is a property of the thymus (Playfair, Purves & Marshall-Clarke, 1971). NZB thymus cells, even quite early in life (3 weeks to 4 months of age), can produce an apparent graft-*versus*-host reaction when injected into newborn syngeneic mice (Playfair & Krsiakova, 1969; however see Cantor, Asofsky & Talal, 1970), and grafting of NZB thymus to mice of other strains results in production of red cell autoantibodies in the recipients (Howie & Helyer, 1966). The situation in the intact NZB mouse may be comparable to a chronic graft-*versus*-host reaction, as suggested by Cole & Nowell (1970). One would then find a progressive loss of immune responsiveness, such as is observed in graft-*versus*-host reactions (Blaese, Martinez & Good, 1964) and the late development of lymphomas

147

characteristic both of old NZB mice (Mellors, 1967; East, 1970) and of chronic graft-*versus*-host disease (Armstrong *et al.*, 1970; Cole & Nowell, 1970). The production of autoantibodies may simply result from the reaction of non-tolerant T cells against auto-antigens such as those of erythrocytes or may be the manifestation of an 'allogeneic effect', in which a graft-*versus*-host reaction serves non-specifically to co-operate in the antibody response to other antigens (Katz *et al.*, 1971). Cantor and his co-workers, studying synergy between different T lymphocyte populations in the production of graft-*versus*-host disease, have demonstrated that only one type of T cell, that which commonly recirculates, may be deficient in old NZB mice (Cantor, Asofsky & Talal, 1970). This implies that there remains a 'normal' T cell subpopulation which must be included among the θ-positive cells which we have enumerated.

It must be added that, while our data and the other studies cited demonstrate the presence in older NZB spleen and lymph nodes of an abnormal T cell population, we have not shown that the initial abnormality is a property of the T cell as such. Several reports suggest that the bone marrow, which provides stem cells to the thymus, may be the seat of the abnormality. Transfer of young NZB bone marrow to newborn CBA mice caused these to become Coombs'-positive 2–3 months later (Allman *et al.*, 1969), and the abnormal tolerance responses of young NZB mice could be conferred on older animals by similar transfer of young NZB marrow (Staples, Steinberg & Talal, 1970).

The second principal finding of the present study is that the reticulum cell sarcoma, which commonly appears in older NZB mice (Mellors, 1967; East, 1970), is negative for both θ and MBLA. This finding is consistent with the non-lymphoid character of the tumour, which presumably bears no relation to the autoantibody formation observed in these mice but may well be a by-product of the immunosuppression which they experience or of reticulum cell hyperplasia accompanying their graft-*versus*-host reaction. The thymoma, which we studied and which proved to be θ-positive, TL-positive and MBLA-negative, is an uncommon tumour in NZB mice (see East, 1970), appears in both Coombs'-positive and Coombs'-negative animals, and therefore cannot be related directly to the immunologic abnormality present in these animals.

ACKNOWLEDGMENTS

The authors express their gratitude to Theresa Killick for technical assistance in carrying out the cytotoxicity measurements and to Dr N. A. Mitchison for his kindness in providing facilities for the carrying out of this work. Dr Waksman was supported by special fellowship of the National Institutes of Health No. Ca-00680 and Dr Raff by a postdoctoral fellowship of the National Multiple Sclerosis Society of the United States.

Dr Waksman was on leave from the Department of Microbiology, Yale University School of Medicine, New Haven, Connecticut, U.S.A.

REFERENCES

ALLMAN, V., GHAFFAR, A., PLAYFAIR, J.H.L. & ROITT, I.M. (1969) Transfer of autoantibody formation by NZB bone marrow cells. *Transplantation*, **8**, 899.
ARMSTRONG, M.Y.K., GLEICHMANN, E., GLEICHMANN, H., BELDOTTI, L., ANDRÉ-SCHWARTZ, J. & SCHWARTZ, R. S. (1970) Chronic allogeneic disease. II. Development of lymphomas. *J. exp. Med.* **132**, 417.

148

BLAESE, R.M., MARTINEZ, C. & GOOD, R.A. (1964) Immunologic incompetence of immunologically runted animals. *J. exp. Med.* **119**, 211.

CANTOR, H., ASOFSKY, R. & TALAL, N. (1970) Synergy among lymphoid cells mediating the graft-versus-host response. I. Synergy in graft-versus-host reactions produced by cells from NZB/B1 mice. *J. exp. Med.* **131**, 223.

CANTOR, H. & ASOFSKY, R. (1970) Synergy among lymphoid cells mediating the graft-versus-host response. II. Synergy in graft-versus-host reactions produced by BALB/c lymphoid cells of differing anatomic origin. *J. exp. Med.* **131**, 235.

CEROTTINI, J.-C., LAMBERT, P.-H. & DIXON, F.J. (1969) Comparison of the immune responsiveness of NZB and NZB $\times$ NZW F_1 hybrid mice with that of other strains of mice. *J. exp. Med.* **130**, 1093.

COHEN, A. & SCHLESINGER, M. (1970) Absorption of guinea-pig serum with agar. *Transplantation*, **10**, 130.

COLE, L.J. & NOWELL, P.C. (1970) Parental-F_1 hybrid bone-marrow chimeras: High incidence of donor type lymphomas. *Proc. Soc. exp. Biol. (N.Y.)*, **134**, 653.

DENMAN, A.M. & DENMAN, E.J. (1970) Depletion of long-lived lymphocytes in old New Zealand black mice. *Clin. exp. Immunol.* **6**, 457.

DENMAN, A.M., DENMAN, E.J. & HOLBOROW, E.J. (1967) Suppression of Coombs-positive hemolytic anemia in NZB mice by antilymphocyte globulin. *Lancet*, **i**, 1084.

DIENER, E. (1966) The immune response in NZB and NZB $\times$ C$_3$H F_1-hybrid mice as measured by the haemolysin plaque technique. *Int. Arch. Allergy*, **30**, 120.

EAST, J. (1970) Immunopathology and neoplasms in New Zealand Black (NZB) and SJL/J mice. *Progr. exp. Tumor Res. (Basel)*, **13**, 84.

HALPERN, B. & FRAY, A. (1969) Déclenchement de l'anémie hémolytique autoimmune chez de jeunes souriceaux NZB par l'administration de Corynebacterium parvum. *Ann Inst. Past.* **117**, 778.

HOWIE, J.B. & HELYER, B.J. (1965) Autoimmune disease in mice. *Ann. N.Y. Acad. Sci.* **124**, 167.

HOWIE, J.B. & HELYER, B.J. (1966) The influence of neonatal thymectomy and thymus grafting on spontaneous auto-immune disease in mice. *Ciba Foundation Symposium on The Thymus* (Ed. by G. E. W. Wolstenholme and R. Porter), Little Brown, Boston, p. 360.

HOWIE, J.B. & HELYER, B.J. (1968) The immunology and pathology of NZB mice. *Advanc. Immunol.* **9**, 215.

KATZ, D.H., PAUL, W.E., GOIDL, E.A. & BENACERRAF, B. (1971) Carrier function in anti-hapten antibody responses. III. Stimulation of antibody synthesis and facilitation of hapten-specific secondary antibody responses by graft-versus-host reactions. *J. exp. Med.* **133**, 169.

LAMBERT, P.H. & DIXON, F.J. (1968) Pathogenesis of the glomerulonephritis of NZB/W mice. *J. exp. Med.* **127**, 507.

LEVENTHAL, B.G. & TALAL, N. (1970) Response of NZB and NZB/NZW spleen cells to mitogenic agents. *J. Immunol.* **104**, 918.

MELLORS, R.C. (1967) Autoimmune and immunoproliferative diseases of NZB/B1 mice and hybrids. *Int. Rev. Exp. Path.* **5**, 217.

MORTON J.I. & SIEGEL, B.V. (1969) Response of NZB mice to foreign antigen and development of auto-immune disease. *J. Reticuloend. Soc.* **6**, 78.

PLAYFAIR, J.H.L. (1968) Strain differences in the immune response of mice. I. The neonatal response to sheep red cells. *Immunology*, **15**, 35.

PLAYFAIR, J.H.L. & KRSIAKOVA, M. (1969) Syngeneic splenomegaly in NZB and BALB/c mice. *Transplantation*, **7**, 443.

PLAYFAIR, J.H.L., PURVES, E.C. & MARSHALL-CLARKE, S. (1971) Abnormal immune responses by NZB thymus cells. *Brit. Soc. Immunol. Meeting, London*, 23 April.

RAFF, M.C. (1969) Theta isoantigen as a marker of thymus-derived lymphocytes in mice. *Nature (Lond.)*, **224**, 378.

RAFF, M.C. (1971) Surface antigenic markers for distinguishing T and B lymphocytes in mice. *Transplantation Rev.*, **6**, 52.

RAFF, M.C., NASE, S. & MITCHISON, N.A. (1971) Mouse specific bone marrow-derived lymphocyte antigen as a marker for thymus-independent lymphocytes. *Nature (Lond.)*, **230**, 50.

RAFF, M.C. & OWEN, J.J.T. (1971) Thymus-derived lymphocytes: Their distribution and role in the development of peripheral lymphoid tissues of the mouse. *Europ. J. Immunol.* **1**, 27.

RAFF, M.C. & WORTIS, H.H. (1970) Thymus dependence of θ-bearing cells in the peripheral lymphoid tissues of mice. *Immunology*, **18**, 931.

Russell, A.S. & Denman, A.M. (1969) Normal induction and maintenance of tolerance in (NZB × NZW) F₁ mice. *Clin. exp. Immunol.* **5,** 265.

Russell, P.J., Hicks, J.D. & Burnet, F.M. (1966) Cyclophosphamide treatment of kidney disease in (NZB × NZW) F₁ mice. *Lancet,* **i,** 1279.

Staples, P.J., Steinberg, A.D. & Talal, N. (1970) Induction of immunologic tolerance in older New Zealand mice repopulated with young spleen, bone marrow, or thymus. *J. exp. Med.* **131,** 1223.

Staples, P.J. & Talal, N. (1969a) Rapid loss of tolerance induced in weanling NZB and B/W F₁ mice. *Science,* **163,** 1215.

Staples, P.J. & Talal, N. (1969b) Relative inability to induce tolerance in adult NZB and NZB/NZW F₁ mice. *J. exp. Med.* **129,** 123.

Staples, P.J. & Talal, N. (1970) Impaired recovery of plaque-forming cells in New Zealand mice given cyclophosphamide. *J. Immunol.* **104,** 1101.

Steinberg, A.D., Baron, S. & Talal, N. (1969) The pathogenesis of autoimmunity in New Zealand mice. I. Induction of antinucleic acid antibodies by polyinosinic-polycytidylic acid. *Proc. nat. Acad. Sci. (Wash.),* **63,** 1102.

Stutman, O., Yunis, E.J. & Good, R.A. (1968) Deficient immunologic functions of NZB mice. *Proc. Soc. exp. Biol. (N.Y.),* **127,** 1204.

Weir, D.M., McBride, W. & Naysmith, J.D. (1968) Immune response to a soluble protein antigen in NZB mice. *Nature (Lond.),* **219,** 1276.

Zatz, M.M., Mellors, R.C. & Lance, E.M. (1971) Changes in lymphoid populations of aging CBA and NZB mice. *Clin. exp. Immunol.* **8,** 491.

Cytotoxic effect of anti-Θ and anti-mouse specific B lymphocyte antigen (MBLA) antisera on helper cells and antibody-forming cell precursors in the immune response of mice to the 4-hydroxy-3,5-dinitrophenacetyl (NNP) hapten

J. Niederhuber , **Erna Möller** and
O. Mäkelä

Abbreviations: **AFCP:** Antibody-forming cell precursors
BSA: Bovine serum albumin   **CG:** Chicken globulin
HRBC: Horse red blood cells   **LHG:** Local hemolysis in gel
MBLA: Mouse specific B lymphocyte antigen   **NNP:** 4-hydroxy-
3,5-dinitrophenacetyl   **NS:** Normal serum   **OA:** Ovalbumin
ORBC: Ox red blood cells   **PFC:** Plaque-forming cells

Anti-Θ antibodies, but not anti-MBLA antibodies, were cytotoxic to helper cells in the adoptive immune response of mice to the hapten, NNP. On the other hand, antibody-forming cell precursors were resistant to anti-Θ antibodies, but sensitive to the cytotoxic effect of anti-MBLA antibodies. This suggests that the use of these reagents allows the purification of viable T and B lymphocytes from peripheral lymphoid organs of mice.

1. Introduction

To analyze the relative roles of T and B lymphoid cell lines in the immune response of mice to different antigens, it is necessary to separate the two as peripheral lymphoid organs contain a certain defined proportion of T and B cells. The spleen contains mainly B cells, whereas lymph nodes are rich in T cells [1].

151

Since the introduction of the Θ alloantigen as a marker for
T lymphoid cells [1], the role of the T cell has been studied
and is now known to mediate cellular immune reactions and
to act as a helper cell in the induction of humoral immune
responses to certain antigens [2, 3]. Furthermore, it has been
established that the immunological memory function is dependent on activated T and B lymphoid cells as well [4–7].

Recently, Raff et al. [8] described the MBLA antigen(s),
which is present on peripheral B lymphoid cells, but not on
T cells. The data suggested that MBLA could be used as a
marker for B cells. We have studied the effect of Θ antiserum
and of MBLA antiserum on the cell populations which are
required for an adoptive immune response to the hapten
NNP in mice.

2. Materials and methods

2.1. Animals

The following F_1 hybrid mice were used in the present study:
A x CBA, A x C57BL, A x C57L and A x B10.5M. AKR mice
were used for the production of anti-Θ C3H serum. Irradiation of mice was carried out in a Siemens X-ray machine at
15 mA and 185 kV; 650 r was given as a whole body irradiation dose, one day before transfer of cells.

Mice were immunized against the hapten, NNP (4-hydroxy-3,5-
dinitrophenacetyl) with $NNP_{12}BSA$ (12 moles of NNP/mole
of BSA) or with $NNP_{17}CG$, and against OA, using alumpreci-
pitated protein injected with 1 x 10^9 pertussis organisms as
an adjuvant. The injections were given 24 − 41 days before
harvest of immune spleen-cells. Immunizations with HRBC
(horse red blood cells) or ORBC (ox red blood cells) were
performed with 4 x 10^8 cells i.p. 6 − 8 days before transfer
of cells.

2.2. Antisera

Anti-MBLA serum was raised in rabbits by injections of
lymphoid cells from T-cell deprived mice, as described by
Raff et al. [8]. The serum was heat-inactivated, absorbed
with sheep red blood cells, mouse red blood cells, mouse
liver and repeatedly with mouse thymus cells. IgG fractions
were prepared on an ion exchange column (DEAE-Sephadex)
as described before for preparation of anti-MBLA serum
[9]. The resultant IgG fraction was not cytotoxic to mouse
thymus cells, but had a cytotoxic effect on bone marrow

cells, spleen cells and lymph node cells, as described before
[10]. The cytotoxic activity of the serum could be absorbed
with spleen, lymph node and bone marrow cells, but not with
thymus cells. The cytotoxic titer of the serum against bone
marrow cells was 1/8 — 1/16, using agarose absorbed guinea
pig complement. The activity of the anti-MBLA serum was
not due to the presence of anti-Ig antibodies [10].

Anti-Θ serum was produced in AKR mice by weekly injections
for 6 — 10 weeks with thymus cells from CBA mice. The serum
was heat-inactivated, but not absorbed before use. It had a
cytotoxic titer of 1/256 against thymus cells with agarose
absorbed guinea pig complement.

2.3. Antiserum treatment of spleen cells

Packed and washed spleen cells were treated *in vitro* with
anti-Θ, anti-MBLA or normal serum diluted 1/2 for 30 min
at 37 $^\circ$C. For 100 x 10^6 spleen cells, 1 — 1.5 ml of serum
was used. Thereafter the cells were washed, complement
added, and the incubation continued for another 30 min.
The cells were again washed, the cytotoxic effect of the antiserum assessed and the cells adjusted to the desired cell concentration for injection. As a rule, 10 x 10^6 — 20 x 10^6 cells
of immune helper cells and of immune hapten-reactive cells
were used. Most groups were challenged with antigen at the
time of transfer (see Tables). Six to seven days after cell transfer, the mice were killed, spleens removed and the number of
anti-hapten antibody-producing cells was estimated with the
local hemolysis in gel assay as described by Jerne and Nordin
[11], with modifications for hapten-coated red cells as described by Pasanen and Mäkelä [12]. NNP-coupled sheep red
blood cells were used as target cells in the LHG assay.

3. Results

The effect of our reagents was first tested on spleen cells
from mice immunized against NNP-CG. Spleen cells were
divided into 4 groups. One group of cells was left untreated,
one was treated with normal serum, one with anti-Θ serum
and another with anti-MBLA serum. Complement was added
to all serum-treated groups. Groups of irradiated syngeneic
mice were injected with the cells as shown in Table 1. NNP-
CG was injected at the time of transfer to groups 1a — 1d.
Six days after transfer, the number of direct and indirect
PFC was determined in the recipients' spleens. It was found
that only group, 1a (given normal serum treated cells), gave a
good response of indirect PFC. The reduction of the immune

response after treatment with anti-Θ or anti-MBLA was more than 90 %. In order to establish that the effect of anti-Θ serum was to reduce the activity of T cells, which are needed for the secondary response to the hapten-protein conjugate, and the effect of anti-MBLA serum to be cytotoxic to the antibody-forming cell precursors (AFCP), experiments were carried out in a test system where the "helper" cells and the AFCP population were derived from different animals, according to the scheme of Mitchison et al. [13].

One group of mice was immunized against HRBC or ORBC to produce helper cells. Another group of mice were primed against the hapten by immunization with NNP-BSA or NNP-CG. At the time of cell transfer the mice were challenged with NNP-coupled HRBC. Experiment nos. 2 and 3 (Table 2) show two tests where the effect of antiserum treatment of helper cells was studied. Spleen cells from mice immunized against HRBC (or ORBC as controls) were treated with normal serum, anti-Θ or anti-MBLA serum plus complement. Thereafter, cells from mice immune to the hapten were added to the helper cells, except in groups f, which were not given hapten-primed cells. The cell mixtures were injected with antigen i.v. and the immune response of irradiated recipient spleens was measured on day 6 (Exp. 2) and on day 7 (Exp. 3). The results showed that anti-Θ serum treatment of helper cells reduced the adoptive immune response with regard to indirect PFC. The number of direct PFC was low, but seemed to be less affected by anti-Θ serum treatment of helper cells. Anti-MBLA serum treatment, however, did not impair the helper cell activity at all.
Experiments 4 and 5, shown in Table 3, were designed to test the activity of anti-Θ and anti-MBLA sera on AFCP. Helper cells were derived from HRBC immune mice.

Cells primed to the hapten were derived from the spleens of NNP-CG or NNP-BSA immunized mice. We found that the memory function of hapten-precursor cells was optimal 20 − 40 days after priming. Cells from mice immune to the hapten were treated with antisera as before, admixed with immune helper cells and the immunogen (NNP-HRBC) and injected into irradiated syngeneic mice. Seven days later the spleens were harvested and direct and indirect PFC reacting with NNP-SRBC were determined. The results indicated that the hapten-reactive precursor cells were resistant to the cytotoxic effect of anti-Θ serum, but were partially sensitive to the action of anti-MBLA antiserum, since only the latter

154

Table 1. Cytotoxic effect of anti-Θ and anti-MBLA on cells prior to adoptive transfer into irradiated syngeneic recipients

Exp. No.	Immunization of cell donor	No. of cells	Serum treatment of cells	Adoptive immune response against NNP	
				Direct PFC per 10^6 cells[a]	Indirect PFC per 10^6 cells[a]
1 a	400 μg NNP-CG[b]	1×10^7	AKR-NS	1.82 ± 0.21 (66)	3.20 ± 0.32 (1598)
1 b	400 μg NNP-CG	1×10^7	Anti-Θ	1.50 ± 0.08 (32)	1.83 ± 0.10 (67)
1 c	400 μg NNP-CG	1×10^7	Anti-MBLA	1.82 ± 0.08 (66)	2.01 ± 0.13 (103)
1 d	none	1×10^7	–	1.98 ± 0.31 (96)	1.12 ± 0.31 (13)
1 e	400 μg NNP-CG	1×10^7	–	1.22 ± 0.11 (17)	0.61 ± 0.06 (4)

a) The results are expressed in $\log_{10}$ for the geometric mean ± standard error; the corresponding anti-logs in parentheses.
b) All groups (except 1 e) were challenged with antigen (400 μg NNP-CG) at the time of transfer.

Table 2. Cytotoxic effect of anti-Θ and anti-MBLA antisera on helper cells

Exp. No.	Immunization of helper cell donor	Serum treatment of helper cells	No. of cells	Additional cells from mice immunized against	No. of cells	Adoptive immune response day 7 Direct PFC per 10^6 cells[a]	Indirect PFC per 10^6 cells[a]
2 a	4×10^8 HRBC	AKR-NS	1.8×10^7	150 μg NNP-BSA	2×10^7	2.01 ± 0.14 (103)	2.52 ± 0.08 (330)
b	4×10^8 HRBC	Anti-Θ	1.3×10^7	150 μg NNP-BSA	2×10^7	1.82 ± 0.05 (66)	2.10 ± 0.05 (126)
c	4×10^8 HRBC	Anti-MBLA	0.5×10^7	150 μg NNP-BSA	2×10^7	2.02 ± 0.08 (105)	2.47 ± 0.08 (294)
d	4×10^8 ORBC	–	2.6×10^7	150 μg NNP-BSA	2×10^7	1.48 ± 0.07 (30)	2.08 ± 0.03 (120)
e	–	–		150 μg NNP-BSA	2×10^7	1.91 ± 0.08 (82)	2.09 ± 0.14 (124)
f	4×10^8 HRBC	–	1.8×10^7	–		1.45 ± 0.17 (28)	1.60 ± 0.17 (39)
3 a	4×10^8 HRBC	AKR-NS	1.9×10^7	150 μg NNP-BSA	2×10^7	1.26 ± 0.19 (18)	2.33 ± 0.14 (211)
b	4×10^8 HRBC	Anti-Θ	1.2×10^7	150 μg NNP-BSA	2×10^7	1.16 ± 0.06 (15)	1.58 ± 0.11 (38)
c	4×10^8 HRBC	Anti-MBLA	1.0×10^7	150 μg NNP-BSA	2×10^7	1.29 ± 0.07 (19)	2.52 ± 0.03 (331)
d	4×10^8 ORBC	–	1.9×10^7	150 μg NNP-BSA	2×10^7	1.37 ± 0.02 (23)	2.10 ± 0.05 (126)
e	–	–		150 μg NNP-BSA	2×10^7	1.44 ± 0.08 (28)	2.17 ± 0.02 (149)
f	4×10^8 HRBC	–	1.9×10^7	–		0.83 ± 0.11 (7)	0.96 ± 0.08 (9)

a) The results are expressed in $\log_{10}$ for the geometric mean ± standard error; the corresponding anti-logs in parentheses.

Table 3. Cytotoxic effect of anti-Θ and anti-MBLA on antibody-forming cells

Exp.	Helper cells from donors immune to	No. of cells	Precursor cells from mice immune to	Serum treatment of precursor cells	No. of cells	Adoptive immune response day 7 Direct PFC per 10^6 cells[a]	Indirect PFC per 10^6 cells[a]
4 a	4×10^8 HRBC	2×10^7	150 μg NNP-BSA	—	1.5×10^7	1.68 ± 0.06 (48)	2.71 ± 0.05 (513)
b	4×10^8 HRBC	2×10^7	150 μg NNP-BSA	Anti-Θ	1.5×10^7	1.60 ± 0.13 (40)	2.62 ± 0.13 (417)
c	4×10^8 HRBC	2×10^7	150 μg NNP-BSA	Anti-MBLA	1.5×10^7	1.24 ± 0.14 (17)	2.37 ± 0.12 (232)
d	150 μg OA	2×10^7	150 μg NNP-BSA	—	1.5×10^7	1.34 ± 0.16 (22)	2.25 ± 0.11 (180)
e	—		150 μg NNP-BSA	—	1.5×10^7	1.08 ± 0.11 (12)	1.35 ± 0.08 (22)
f	4×10^8 HRBC	2×10^7	—	—		1.10 ± 0.16 (13)	1.77 ± 0.13 (58)
5 a	4×10^7 HRBC	10^7	400 μg NNP-CG	—	10^7	2.40 ± 0.17 (252)	3.50 ± 0.03 (3162)
b	4×10^7 HRBC	10^7	400 μg NNP-CG	AKR-NS	10^7	2.55 ± 0.06 (355)	3.46 ± 0.10 (2884)
c	4×10^7 HRBC	10^7	400 μg NNP-CG	Anti-Θ	10^7	2.28 ± 0.20 (191)	3.55 ± 0.06 (3548)
d	4×10^7 HRBC	10^7	400 μg NNP-CG	Anti-MBLA	10^7	1.97 ± 0.07 (93)	3.20 ± 0.11 (1585)
e	4×10^7 ORBC	10^7	400 μg NNP-CG	—	10^7	1.69 ± 0.14 (49)	2.30 ± 0.12 (200)
f	—		400 μg NNP-CG	—	10^7	1.88 ± 0.14 (76)	2.59 ± 0.11 (390)
g	4×10^7 HRBC	10^7	—	—		2.32 ± 0.08 (209)	2.88 ± 0.16 (759)

a) The results are expressed in $\log_{10}$ for the geometric mean ± standard error; the corresponding anti-logs in parentheses.

groups had reduced numbers of direct and indirect plaque-forming cells against NNP. The response of the anti-MBLA serum on primed precursor cells is not complete, however, since the response of groups 4c and 5d with regard to indirect PFC is higher than the response of groups 4f and 5g, given helper cells alone. It should be noted that the expected response should not be extremely low, since non-antiserum treated helper cells, containing non-primed AFCR were present in all groups treated with antisera. However, the reduction of the antibody-forming cell capacity of those spleens is clear, if corrections are made for the response of helper cells alone, and suggests that anti-MBLA antiserum, but not anti-Θ serum, has a cytotoxic effect on AFCP.

4. Discussion

It has been demonstrated earlier that anti-Θ antibodies have a selective cytotoxic effect on thymus-derived lymphocytes. Anti-Θ antibodies have been found to inhibit helper cell activity, and to inhibit, partly or completely, the memory effect in an adoptive transfer test to thymus-dependent antigens. Furthermore, anti-Θ serum does not have any effect on antibody-producing cells [10] or on AFCP [14]. The effect of anti-Θ treatment of spleen cells from mice immune to SRBC can be completely restored in an adoptive response by the addition of anti-MBLA [10] or anti-Ig serum treated [7] spleen cells from the same donor. These results indicate that anti-Θ reacts with one cell population, anti-MBLA and anti-Ig serum with another cell population. Further proof that anti-Θ and anti-MBLA antisera react with different cell populations came from experiments with double immunofluorescent staining, which demonstrated that MBLA was present only on immunoglobulin-bearing B lymphocytes in the mouse. The percentages of thymus-dependent and thymus-independent cells detected by indirect staining of specific antisera treated cells was consistent with the cytotoxic indices for these sera. The double staining of anti-MBLA-treated cells following incubation with fluorescein conjugated anti-mouse immunoglobulin serum and rhodamine conjugated anti-rabbit immunoglobulin serum confirmed the absence of anti-immunoglobulin activity in the anti-MBLA serum*.

---

* Niederhuber, J., Britton, S. and Bergquist, R., submitted for publication.

In the present study we established that anti-Θ serum abolished the effect mediated by primed helper cells, whereas it did not interfere with the activity of primed AFCP. Anti-MBLA antibodies, on the other hand, did not interfere with helper cell activity, but induced a decrease in the response of precursor cells from primed mice, thus indicating that it had an effect on the antibody-forming cell precursors. In some experiments, the effect on AFCP was not complete, indicating that few precursors escaped cytotoxicity. However, as reported earlier, anti-MBLA does kill the large majority of antibody-secreting cells [10], known to be of B cell origin [15].

From these studies we conclude that helper cells are T cells, sensitive to anti-Θ antibodies but resistant to anti-MBLA antibodies and that antibody-forming cell precursors are resistant to anti-Θ antibodies, but are sensitive to the cytotoxic effect of anti-MBLA antibodies plus complement. The two reagents react with different cell populations (T and B cells, respectively) and can be used to purify these two cell populations from peripheral lymphoid organs. Such purified populations of cells are viable and give rise to the expected responses upon transfer into irradiated recipients.

The skillful technical assistance of Mrs. Lill-Britt Andersson is gratefully acknowledged. The present work was supported by grants from the Swedish Medical Research Council, the Wallenberg Foundation and the Harald Jeansson Foundation.

Received March 7, 1972; in revised form May 9, 1972.

5. References

1 Raff, M.C., *Nature* 1969. *224*: 378.

2 Raff, M.C., *Nature* 1970. *226*: 1257.

3 Schimpl, A. and Wecker, E., *Nature* 1970. *226*: 1258.

4 Takahashi, T., Carswell, E.A. and Thorbecke, G.J., *J. Exp. Med.* 1970. *132*: 1181.

5 Möller, E. and Greaves, M.F., in Cross, A.M., Kosunen, T.U. and Mäkelä, O. (Eds.) *"Cell Interactions in Immune Responses"*, 3rd Sigrid Juselius Symposium, Academic Press, New York 1970, p. 101.

6 Niederhuber, J. and Möller, E., *Cell. Immunol.* 1972, in press.

7 Takahashi, T., Mond, J.J., Carswell, E.A. and Thorbecke, G.J., *J. Immunol.* 1971. *107*: 1520.

8 Raff, M.C., Nase, S. and Mitchison, N.A., *Nature* 1971. *230*: 50.

9 Niederhuber, J., *Nature* 1971. *233*: 86.

10 Niederhuber, J. and Möller, E., *Cell. Immunol.* 1972. *3*: 599.

11 Jerne, N.K. and Nordin, A.A., *Science* 1965. *140*: 405.

12 Pasanen, V.J. and Mäkelä, O., *Immunology* 1969. *16*: 399.

13 Mitchison, N.A., Rajewsky, K. and Taylor, R.B., in Sterzl, J., and Riha, I. (Eds.), *Prague Symposium on Developmental Aspects of Antibody Formation and Structure*, Publishing House of the Czechoslovak Academy of Sciences, Prague 1970, p. 547.

14 Niederhuber, J. and Möller, E., *Cell. Immunol.* 1972, in press.

15 Mitchell, G.F. and Miller, J.F.A.P., *J. Exp. Med.* 1968. *128*: 821.

RECEPTOR FOR ANTIBODY-ANTIGEN
COMPLEXES USED TO SEPARATE
T CELLS FROM B CELLS

A . BASTEN, J . SPRENT and J. F. A. P. MILLER

LYMPHOCYTES with the capacity to bind antibody–antigen complexes to their surface[1-3] are probably bone marrow-derived, B, cells, not thymus-derived, T, cells[3]. We now have definite evidence that such lymphocytes are indeed B cells, and will describe how this property can be utilized in a practical way for separating T cells from B cells.

The presence of immune complexes on the cell surface was demonstrated by an autoradiographic technique. Fowl immunoglobulin (FγG), prepared as described previously[4] and absorbed extensively against mouse lymphocytes to remove nonspecific anti-mouse activity, was iodinated[5] with ^{125}I (Radiochemical Centre, Amersham, catalogue number IMS3). Both levels of specific activity (50 and 150 µCi/µg FγG) used gave the same number of labelled cells if exposure times were adjusted appropriately. Lymphocytes were obtained by thoracic duct cannulation[6] of highly inbred CBA mice. After washing, cell suspensions were incubated with inactivated hyperimmune mouse anti-FγG serum or normal mouse serum (NMS) for 30 min at 37° C. Excess serum was removed by centrifugation and the cells resuspended in Dulbecco's balanced salt solution containing ^{125}I-FγG for a further 30 min in the cold[5]. Smears for autoradiography were dipped in Kodak NTB-2 emulsion and developed 2–10 days later.

As Table 1 shows, 12–20% of thoracic duct lymphocytes (TDL) from normal mice bound the immune complex. Labelling was inhibited by preincubation of antibody-coated cells with excess ^{125}I-FγG. Washing the TDL before reaction with radioactive antigen reduced the number of labelled cells to <1%. By contrast, no diminution in labelling was observed when the cells were washed two to four times after incubation with ^{125}I-FγG. These findings imply that the bond between cell and antibody is weak but that the formation of an antibody–antigen complex on the cell surface stabilizes the bond. The number of TDL binding ^{125}I-FγG rose with increasing anti-

Table 1 Binding of Antibody–Antigen Complexes to Thoracic Duct Lymphocytes

Serum used for pretreatment of TDL	% Labelled with	
	^{125}I-FγG *	^{125}I-HGG †
Normal	<1	<1
Anti-FγG	17 (12–20)	<1
Anti-HGG	<1	15 (12–19)

The same percentage of cells labelled in the presence of EDTA.

* Figures obtained from six experiments.
† Figures obtained from two experiments.

body levels for titres ranging from 1 : 2 to 1 : 16. When, however, antisera of higher titres were used, no further increase in labelled cells was observed, the mean peak value being 17% (Table 1). This strongly suggested that only a subpopulation of lymphocytes could bind immune complexes. The capacity of other complexes such as anti-HGG–HGG (HGG, human gamma globulin) and anti-POL–POL (POL, polymerized flagellin of *Salmonella adelaide*) to produce a similar effect established the nonspecificity of the phenomenon.

Mouse TDL are composed of T and B cells; 75–85% have the characteristics of T cells in that they are susceptible to lysis by anti-θ C3H serum and complement[7]. Thus, the minority population of lymphocytes binding antibody–antigen complexes (12–20%) might be composed of B cells or a subclass of T cells. This problem was clarified by using populations of lymphocytes enriched for either T or B cells. Thoracic duct cannulation of athymic (Nude-nu nu) mice[8] provided a source of B cells (B.TDL). A pure population of T cells was obtained by the method of Sprent and Miller[9]. Lethally irradiated

162

Table 2 Selective Binding of Antibody–Antigen Complexes to B Cells *

Type of TDL	% Labelled with anti-FγG-^{125}I-FγG †	% Killed by anti-θ C3H serum †
Normal	16 (13–18)	79 (78–83)
B.TDL ‡	97 (95–98)	0
T.TDL	0.3 (0.1–0.5)	90 (87–94)
TDL collected after thoracic duct drainage for 8 days	45 (42–46)	50 (43–56)

* Autoradiographic and cytotoxic studies were performed with the same batches of anti-FγG serum and anti-θ C3H serum.

† Figures obtained from two to three experiments.

‡ Obtained from athymic (nu nu) mice; nu$^+$ mice were shown to possess the θ C3H allele.

Table 3 Depletion of B Cells by Passage of Antibody Coated TDL through a Column of Antigen Coated Beads

Source of TDL	% Labelled with anti-FγG-^{125}I-FγG *	% Killed by anti-θ C3H serum *
Normal	17 (15–19)	80 (75–84)
Column effluent after NRS treatment	15 (11–20)	82 (80–84)
Column effluent after anti-HGG treatment	1.3 (0.5–2.2) †	91 (90–94)

Total cell recovery was 70–75% after NRS treatment and 60% after anti-HGG treatment.

* Figures obtained from three experiments.

† The mean grain count of the cells in this group was thirty-five compared with > 100 for controls. The only B cells that escaped thus appear to be those with receptors of low density.

(CBA × C57BL)F$_1$ mice were given 10^8 CBA thymus cells and cannulated 4 days later. The TDL obtained were identified with anti-H2 sera as CBA in origin and were therefore derived from the original thymus cell inoculum. They will be referred to as T.TDL. Treatment of these cell populations with anti-θ C3H serum and complement killed approximately 90% of T.TDL but no B.TDL (Table 2). Both types of TDL were also reacted with NMS or anti-FγG serum and then ^{125}I-FγG as described above. The autoradiographic findings are shown in Table 2 where they are compared with those for TDL from normal mice. Between 95 and 98% of B.TDL bound the immune complex, whereas less than 1% of T.TDL did so (even after exposure of the smears for 60 days). Furthermore, T cell depletion of the recirculating lymphocyte pool by prolonged thoracic duct drainage increased the proportion of labelled lymphocytes by three to four times. B cells, not T cells, therefore seem to carry a receptor for antibody–antigen complexes

Table 4 Adoptive Secondary Response of Spleen Cells passed through Antigen Coated Columns

Type of cells *	No. of irradiated recipients	Peak 7S p.f.c. per spleen Anti-FγG (±s.e.)	Anti-HRBC (±s.e.)
Control spleen cells†	6	254,255 (304,590–212,235)	274,285 (300,245–250,570)
B depleted spleen cells‡	4	1,325 (2,225–790)	910 (1,550–540)
B depleted spleen cells‡ + primed B cells§	5	129,175 (146,085–114,225)	49,325 (61,490–39,565)
Primed B cells§	6	1,415 (2,205–905)	2,360 (3,005–1,855)
B depleted spleen cells† + normal TDL	5	75 (445–10)	7,070 (12,150–4,115)
Normal TDL	8	65 (85–50)	1,610 (1,885–1,375)
Spleen cells eluted from column	6	375 (1,385–100)	450 (1,610–125)
Spleen cells eluted from column + normal TDL	6	111,045 (155,250–79,425)	177,205 (208,060–150,925)

Total cell recovery was 50–55% after NRS treatment and 30–35% after anti-HGG treatment.
* Number of viable cells of each kind per recipient, 5×10^6.
† Spleen cells incubated in NRS and passed through HGG coated column.
‡ Spleen cells incubated in anti-HGG serum and passed through HGG coated column.
§ Obtained by treating spleen cells from mice primed to FγG and HRBC with anti-θ C3H serum and complement.

164

on their surface. Preliminary studies with rat TDL and mono-nuclear cells from human peripheral blood indicate that a sub-population (15–35%) of these cells likewise bind immune complexes. Experiments to be reported elsewhere suggested that complement was not required for the binding of soluble complexes to B cells (unpublished results of A. B.).

The existence of a receptor of this kind on B cells permitted development of a method for depleting B cells from lymphoid cell populations. Polymethylmethacrylic plastic beads (Dega-lan) were coated with HGG according to the method of Wigzell and Andersson[10]. Trace labelling studies with ^{125}I-HGG revealed that the optimal concentration of HGG for coating was 10 mg/ml. After washing in sterile saline, the beads were poured into 1×15 cm glass columns and equilibrated with medium 199 containing 10% NMS for 1–2 h at 4° C. Lymphocytes from thoracic duct lymph or spleen were incubated with normal rabbit serum (NRS) or rabbit anti-HGG serum (5×10^7 cells/ml. rabbit serum diluted 1 : 5) at 37° C for 30 min. The cells were then centrifuged once, resuspended in medium 199 and approximately 15×10^7 cells run into each column of beads. The columns were then shut off and held horizontally for a further hour at 4° C. Subsequently, lymphocytes were eluted with the columns in the vertical position. Their viability as measured by trypan blue exclusion was always greater than 98% in the case of TDL and 70–80% for spleen cells. It was predicted that B cells in those populations incubated with anti-HGG sera would be retained on the antigen coated beads and T cells excluded. This was verified by examining the markers present on the cells in the column effluent. Table 3 shows that the emerging cells were susceptible to anti-θ serum and complement although only 1 in 100 bound an immune complex such as anti-FγG-^{125}I-FγG. By contrast, lymphocytes incubated with NRS and passed through antigen coated beads contained the same proportion of T and B cells present in the original population.

An independent assessment of the validity of the method was obtained by studying the immune capacity of the emerging cells in an adoptive transfer system. For this purpose, spleen cells from mice primed to FγG and horse erythrocytes (HRBC) were incubated with anti-HGG serum or NRS and passed through a column as before. Cells from the effluent were injected into heavily irradiated (750 rad) mice together with fluid FγG and HRBC (Table 4). As expected, lymphocytes incubated with NRS elicited an excellent secondary response. Irradiated recipients of anti-HGG treated cells, however, failed to do so. To check whether B cells had been retained on the column, anti-HGG treated cells were transferred with either T cells (from normal TDL which do not respond to fluid FγG)[11] or B cells. The latter were obtained by treating spleen cells from FγG and HRBC primed mice with anti-θ serum and complement. The results (Table 4) show that B cells but not T cells were able to collaborate with the cells emerging from the column. The cells retained on the column were removed by shaking the antigen coated beads in excess medium 199. On injection into irradiated hosts, they likewise were unable to respond on their own, but in this case their capacity to transfer a response was restored by addition of T cells, not B cells. Taken together, these findings strongly suggest that B cells are selectively retained and that the technique provides a reliable way of producing a cell population greatly enriched for T cells.

We thank Miss Janet Irwin, Miss Ludmilla Ptschelinzew and Miss Jill Phillips for technical assistance. A. B. is in receipt of a Queen Elizabeth II fellowship. The investigations were supported by the National Health and Medical Research Council of Australia, the Australian Research Grants Committee, the Damon Runyon Memorial Fund for Cancer Research, the British Heart Foundation and the National Heart Foundation of Australia.

[1] Uhr, J. W., and Phillips, J. M., *Ann. NY Acad. Sci.*, **129**, 793 (1966).
[2] Bianco, C., Patrick, R., and Nussenzweig, V., *J. Exp. Med.*, **132**, 702 (1970).
[3] Dukor, P., Bianco, C., and Nussenweig, V., *Proc. US Nat. Acad. Sci.*, **67**, 991 (1970).
[4] Miller, J. F. A. P., and Warner, N. L., *Intern. Arch. Allergy Appl. Immunol.*, **40**, 59 (1971).
[5] Byrt, P., and Ada, G. L., *Immunology*, **17**, 503 (1969).
[6] Miller, J. F. A. P., and Mitchell, G. F., *J. Exp. Med.*, **128**, 801 (1968).
[7] Miller, J. F. A. P., and Sprent, J., *Nature*, **230**, 267 (1971).
[8] Pantelouris, E. M., *Nature*, **217**, 370 (1968).
[9] Sprent, J., and Miller, J. F. A. P., *Cell. Immunol.* (in the press).
[10] Wigzell, H., and Andersson, B., *J. Exp. Med.*, **129**, 23 (1969).
[11] Miller, J. F. A. P., and Sprent, J., *J. Exp. Med.*, **134**, 66 (1971).

Quantification of Antigen Specific T and B Lymphocytes in Mouse Spleens

G. E. ROELANTS

SPECIFIC T and B lymphocytes can be killed by radioactive antigen indicating that both B and T cells must have antigen specific receptors[1,2]. Thus, the labelled cells observed by autoradiography after incubation of lymphocytes with radioactive antigen of high specific activity (reviewed in ref. 3) should represent a mixture of specific T and B cells.

The aim of these experiments was to quantify the number of T and B lymphocytes specific for *Maia squinado* haemocyanin (MSH)[2] or extensively iodinated poly (Tyr, Glu)-poly-DL-Ala-poly-Lys type 509 (TIGAL)[4] in spleens of normal and primed mice. This was done by counting the absolute number of labelled cells suppressed by treatment with anti-θ serum and complement. The antibody response to MSH is completely thymus-dependent (unpublished results of G. E. R., A. J. S. Davies, B. A. Askonas and E. B. Jacobson) while it was shown for TIGAL-52, but not yet for TIGAL-509, that only the 7S response is dependent (personal communication from J. H. Humphrey and N. Willcox).

Three types of spleen cells of C_3H mice were used: one from normal mice; one from mice 1 or 6 months after one intraperitoneal injection of 100 μg alum-precipitated MSH and 10^9 *Bordetella pertussis* organisms; and one from mice 3 or 6 months after one intraperitoneal injection of 200 μg alum-precipitated TIGAL and 10^9 *Bordetella pertussis* organisms. The spleen cell suspensions were prepared[2] and divided into two parts: half of the cells were treated with anti-θ serum, and the other half with normal AKR serum, and complement (as described in ref. 5). The anti-θ serum was donated by Dr M. Raff and its specificity against T lymphocytes had been fully established. The cells were then exposed to ^{125}I-MSH or ^{125}I-TIGAL[2]. The specific activities of the antigens ranged from 130 to 400 μCi/μg. The proportion of cells destroyed by anti-θ and complement was determined in each experiment by counting the number of living cells recovered

167

by comparison with the controls: the specific killing varied from 29 to 45%. Autoradiography using 'Ilford K5' emulsion was performed on smears as described[6]. The autoradiographs were developed after varying time intervals (24 h to 10 days incubation), fixed and stained with Giemsa.

For each group from 44,000 to 152,000 lymphocytes were

Table 1 Suppression of MSH Specific Spleen Lymphocytes by Anti-θ Serum and Complement

Spleen lymphocytes	No. per 1,000	> 50 molecules per cell Suppression by anti-θ + complement	No. per 1,000	10–50 molecules per cell Suppression by anti-θ + complement
Unprimed	0.8	NS	1.1	42%
Unprimed	0.8	NS	2.0	42%
Primed (1 month)	6.9	55%	1.7	> 99%
Primed (6 months)	1.1	39%	1.0	> 99%

NS, no significant suppression.

Table 2 Suppression of MSH-specific 6-Month Primed Spleen Lymphocytes by Anti-θ Serum and Complement

Molecules bound per cells	No. per 1,000	Suppression by anti-θ and complement
10–50	1.0	> 99%
> 50	1.1	39%
> 250	0.6	45%
> 750	0.2	NS

NS, no significant suppression.

individually examined in the light microscope by systematically observing adjacent fields at a magnification of 1,000 until all the fields of the smear had been covered. More than 95% of the labelled cells had the morphology of small lymphocytes; this is in agreement with the findings of others[3,6]. Even after treatment with anti-θ and complement few dead cells were seen on the smears, probably due to the passage of the cell suspension through foetal calf serum gradient after exposure to the radioactive antigen[2]. The total number of cells examined was recorded, together with the number of labelled cells; when not confluent, the number of grains at the surface of the labelled cells were counted to estimate the number of molecules of antigen on the surface of the lymphocytes[7]. With 'Ilford K5' emulsion, one disintegration produces 0.4 grain. MSH was degraded after heavy iodination and the molecular weight of the material still precipitable by antibody was ranging from 200,000 to 10^6. The lowest (200,000) was taken for the calculation of molecules per lymphocytes. TIGAL was not degraded and its molecular weight was 232,000. Thus, for a specific activity of 300 μCi/μg, for example, and 7 days of autoradiograph exposure, 10 molecules of MSH would have given 10 grains and 10 molecules of TIGAL 11 grains.

Table 1 presents the results obtained with ^{125}I-MSH. With unprimed spleen cells treatment with anti-θ and complement removed about 40% of the lymphocytes carrying 10 to 50 MSH molecules on their surface, but no decrease in the number of lymphocytes bearing more than 50 MSH molecules could be detected. One month after priming, the number of lymphocytes with more than 50 MSH molecules increased about nine-fold and 55% of them could be destroyed by anti-θ and complement. All the lymphocytes with 10 to 50 MSH molecules were sensitive to anti-θ and complement. The specific activity and duration of autoradiographic exposure were such as to allow a larger distribution of classes when spleen cells were examined 6 months after priming (Table 2). The number of labelled cells in these spleens did not differ significantly from the number found in normal spleens, but 39% of the cells carrying more than 50 MSH molecules and 45% of those with more than 250 molecules were T lymphocytes. Anti-θ did not destroy any of the cells bearing more than 750 MSH molecules representing about 10% of the total population of labelled lymphocytes. In both normal and primed spleens, the maximum number of molecules found on a lymphocyte was estimated at about 5,000.

Table 3 shows the results obtained with ^{125}I-TIGAL. Only primed spleen cells were examined and cells with less than 10 molecules would not have been detected. Most of the labelled lymphocytes had more than 50 TIGAL molecules on their surface; anti-θ and complement destroyed approximately 20% of them. However, cells with more than 150 TIGAL molecules were not destroyed. Here again 5,000 TIGAL molecules was the maximum found on a lymphocyte. Computed from these results, Tables 4 and 5 show the number of MSH and TIGAL specific B and T spleen lymphocytes.

Several observations can be made. B lymphocytes are able to fix more antigen at their surface than T lymphocytes. T cells with more than 50 MSH molecules were not detected in unprimed spleens whereas about 50% of the specific B cells had more than 50 molecules and could have as many as 5,000 (Table 4). In primed spleen no MSH-T lymphocyte was detected with more than 750 molecules; 17% of the equivalent B lymphocytes were binding between 750 and 5,000 MSH molecules (Table 4). The findings with TIGAL primed cells paralleled those with MSH: all labelled T cells had less than 150 molecules, but about one third of the TIGAL-B cells could fix between 150 and 5,000 TIGAL molecules (Table 5). These observations that T cells have fewer receptors than B cells are in agreement with the finding that methods which readily detect immunoglobulin receptors on B lymphocytes fail to do so on T lymphocytes[8]. After priming with MSH,

169

Table 3 Suppression of TIGAL-specific Spleen Lymphocytes by Anti-θ Serum and Complement

Spleen lymphocytes	>150 molecules per cell		>50 molecules per cell		10–50 molecules per cell	
	No. per 1,000	Suppression by anti-θ+complement	No. per 1,000	Suppression by anti-θ+complement	No. per 1,000	Suppression by anti-θ+complement
Primed (3 months)	—	—	5.9	22%	0.8	43%
Primed (6 months)	1.2	NS	3.0	17%	—	—

NS, no significant suppression.

Table 4 MSH-specific B and T Lymphocytes in Spleen

Spleen lymphocytes	No. B lymphocytes per 1,000 Binding per cell				No. T lymphocytes per 1,000 Binding per cell			
	>750 molecules	>250 molecules	>50 molecules	10–50 molecules	>750 molecules	>250 molecules	>50 molecules	10–50 molecules
Unprimed	—	—	0.8	0.6	—	—	0	0.5
Unprimed	—	—	0.8	1.2	—	—	0	0.8
Primed (1 month)	—	—	3.1	0	—	—	3.8	1.7
Primed (6 months)	0.2	0.3	0.7	0	0	0.3	0.4	1.0

Table 5 TIGAL-specific B and T Lymphocytes in Spleen

Spleen lymphocytes	No. B lymphocytes per 1,000 Binding per cell			No. T lymphocytes per 1,000 Binding per cell		
	>150 molecules	>50 molecules	10–50 molecules	>150 molecules	>50 molecules	10–50 molecules
Primed (3 months)	—	4.6	0.5	—	1.3	0.3
Primed (6 months)	1.2	2.5	—	0	0.5	—

both MSH specific B and T lymphocyte populations show a shift towards cells able to fix more antigen at their surface, thus presumably better fitted to recognize MSH. Whether this is due to the appearance or the selection of lymphocytes carrying a larger number of receptors or receptors of higher affinity cannot be decided at present. It should be noted that although a large excess of T over B activity was detected in adoptive transfer experiments using MSH primed spleens[2], the actual number of T and B lymphocytes is approximately equal.

It is difficult to ascertain which cells are effectively involved in the immune induction. Ada has suggested that the relevant cells would be those occurring at a frequency range of 10^{-4} and less[3]. Since the response is suppressed by radioactive antigen, one would expect that only the cells fixing a large number of antigen molecules would be important, although little is known about the amount of ^{125}I needed at the surface of the lymphocyte to kill it[9]. In this respect, the fact that T cells, bearing more than 50 molecules of MSH in unprimed spleens, could not be detected poses a problem. Maybe T lymphocytes fixing more antigen exist in that population, but in too small a quantity to be detected by this method.

It should be emphasized that the numbers given here are valid only for the spleen and do not necessarily reflect the total complement of specific T and B cells of the whole mouse. Nevertheless, it is interesting to note that in adoptive responses with MSH primed spleens[2] the intraperitoneal transfer of 15,000 MSH specific B cells gave a plateau antibody response and that of 5,000 MSH specific T cells a plateau helper effect for DNP-primed cells. Since the proportion of injected cells active in lymph nodes or spleen of a recipient mouse has been estimated to be about 1–5% (ref. 10 and personal communication from J. M. C. Mitchell) 150–750 B and 50–250 T active lymphocytes would be sufficient to start a maximum adoptive response.

I thank Dr B. A. Askonas, Dr J. H. Humphrey and Dr N. Willcox for ^{125}I-MSH degradation analysis.

172

Electrophoretic Mobilities of
T and B Cells

M. WIOLAND
D. SABOLOVIC
C. BURG

IT is now accepted that two classes of lymphocytes are present within the peripheral lymphoid tissues of the mammal: thymus-derived lymphocytes (T cells) and thymus-independent, bone-marrow-derived lymphocytes (B cells). So far much use has been made of isoantigenic markers on T cells to differentiate them from B cells[1-5]; markers which can be used on B cells are more recent[6].

The biophysical characteristics chosen here to distinguish these two types of lymphocytes are those obtained by cell electrophoretic mobility analysis. Our experimental systems are basically the same as those which have been used by others to study the immunological functions of the T and B cells.

We used nude mice bred in sterile conditions. The nude breeders were provided by the Institute of Animal Genetics in Edinburgh. Male and female C57Br and C57Bl/6 mice aged from 2 to 7 months were used, but preliminary experiments showed that neither age nor sex influences electrophoretic mobility. Cell suspensions were obtained from either two or ten animals according to the organ examined. The organs (thymuses and spleens) were dissected immediately after killing; care was taken to separate the thymus from underlying lymph nodes. The organs were disrupted gently in Poter No. 10 (Verrerie Soufflée pour Chimie Industrielle, Paris). The lymphoid cells were suspended in saline and filtered for 30 min at 37° C through a 10 ml. plastic syringe filled with 500 mg of cotton wool to retain plasmocytes and granulocytes but not monocytes and lymphocytes. Different morphological counts made after filtration showed that 95% of the lymphocytes were recovered regardless of size. Two per cent of the granulocytes failed to adhere to the column, but as these were easily distinguishable from the other nucleated cells we did not measure them when taking electrophoretic mobility readings. Whenever necessary, an osmotic shock in distilled water was applied for 15 s to eliminate red blood cells. Finally, the cells were washed three times in saline.

Electrophoretic mobility was determined using a cylindrical microelectrophoresis apparatus[18]. The mobility of human washed erythrocytes was determined before and after each experiment as a control. All measurements were carried out in physiological saline adjusted to pH 7.2 with sodium hydrocarbonate 0.145 M. The mobilities were determined at 25° C $\pm 0.1^\circ$ C and the mean electrophoresis mobility values $\pm$ standard error are expressed in μm s^{-1} V^{-1} cm^{-1}. The numbers of cells scored are given in the tables. The electrophoresis mobility distributions of the cells are summarized by histograms which were decomposed into Gaussian populations using the technique described by Ruhenstroth-Bauer and Lucke-Huhle[19].

To determine the electrophoretic mobility of the B cells, we used the natural system where a population of only B cells is to be found, that is, the homozygous nude mutant described by Flanagan[7], which is a congenital athymic mouse[8] whose spleen has been shown to be deficient in T cells[9]. Most of the lymphoid cells have the surface characteristics of B cells, that is, they are θ negative, MBLA positive[9,10]. Heterozygous littermates, in which B and T cells are present together, were used as control animals. We examined the spleens of 3 homozygous nude mice in 3 separate experiments to determine their mobility. Fig. 1 shows that, as far as the nude homozygous mice are concerned, the electrophoresis mobility of the cell population is 0.81 ± 0.01 μm s^{-1} V^{-1} cm^{-1}. In the case of the nude heterozygous animals (with thymus), the spleen lymphoid cells follow a trimodal pattern. The first class has an electrophoresis mobility of 0.81 μm s^{-1} V^{-1} cm^{-1}; the second, one of 1.20 μm s^{-1} V^{-1} cm^{-1}; and the third, one of 1.37 μm s^{-1} V^{-1} cm^{-1}. In these control animals, the cells with the lowest mobility may be considered as being B cells, the two cell populations of higher mobility as T cells.

For T cells we used Andersson and Blomgren's method[11,12], which showed that mature T cells can be enriched in the thymus at the expense of immature cells by subjecting the mice to a treatment with cortisone. This treatment *in vivo* causes considerable cellular depletion in the thymus: the total number of cells is reduced to about 3 % of its original value. The remaining cells are known as cortisone-resistant thymic cells, and are very efficient in graft-versus-host reactions, responsive to phytohaemagglutinin[13] and efficient as "helper" cells for humoral antibody formation in thymus-dependent systems[12,14]. To determine the electrophoretic mobility of these cortisone-resistant thymic cells, we used heterozygous nude mice (with thymus), and C57Br mice, which had been injected 2 days before with 125 mg/kg of hydrocortisone suspension intraperitoneally. Table 1 shows that there are clear differences between the mean electrophoretic mobility of the thymocyte of a normal mouse and the cortisone-resistant thymic cell.

Table 1 suggests that the electrophoretic mobility of the same cell type may vary according to the mouse strain. We confirmed this by measuring the mobility of the thymocyte in three strains of mice (Table 2).

Our results suggest that the electrophoretic mobility of the thymocyte is strain-dependent. The difference between the values obtained for the cortisone-resistant thymic cell in the C57Br and heterozygous nude mice (Table 1) can be explained on the same basis.

Another method of obtaining purified T cells is that of Mitchell and Miller, who incubated thymus cells with antigen

Table 1 Electrophoretic Mobility of the Thymocyte in Normal Mice and in Cortisone-treated Mice

Cell type	Nude heterozygous	C57Br
Thymocyte in normal mice *	0.71 ± 0.01 (580)	0.77 ± 0.01 (643)
Thymocyte in cortisone treated mice † (Cortisone resistant thymic cell)	1.21 ± 0.01 (120)	1.04 ± 0.03 (263)

The numbers of cells scored are indicated in parentheses.
* Each value represents the mean ± s.e. obtained from ten experiments.
† Each value represents the mean ± s.e. obtained from three experiments for the heterozygous mice and from seven experiments for the C57Br mice.

Table 2 Electrophoretic Mobility Values of the Thymocyte in Three Strains of Mice

C57Br	0.77 ± 0.01 μm s^{-1} V^{-1} cm^{-1}
C57Bl/6	0.84 ± 0.01 μm s^{-1} V^{-1} cm^{-1}
Nude heterozygous	0.71 ± 0.01 μm s^{-1} V^{-1} cm^{-1}

in irradiated mice[15,16] to prepare "educated T cells", so called because the T cells had been in contact with the antigen. We determined their mobility by irradiating ten C57Br mice with 900 rad and injected them intravenously with syngeneic thymus cells mixed with sheep red blood cells (3×10^7 thymocytes + 5×10^8 sheep red blood cells). Eight days later, suspensions of spleen cells were prepared and the electrophoretic mobility of the lymphoid cells measured. This experiment was repeated three times, and in each case the mobility of the "educated T cells" was 1.26 μm s^{-1} V^{-1} cm^{-1}. Thus, the "educated T cell" can be considered to differ from the thymocyte and the cortisone-resistant thymic cell.

In further experiments, we studied more precisely the lymphoid cells present in the spleens of normal mice, using C57Br strain and heterozygous nude mice. The Gaussian distribution of the histograms revealed three cell populations.

In the case of the C57Br strain, the mean electrophoretic

mobility values of the three populations are: 0.71, 1.10 and 1.27 μm s^{-1} V^{-1} cm^{-1}. The results which we obtained earlier by comparing the electrophoretic mobility distribution of spleen cells in homozygous and heterozygous nude mice (Fig. 1) suggest that the cells with the lowest mobility (0.71 μm s^{-1} V^{-1} cm^{-1}) are B cells. The numerical values 1.10 and 1.27 μm s^{-1} V^{-1} cm^{-1} are close to those obtained for the cortisone resistant thymic cell and for the "educated T cell".

In the case of the heterozygous nude mice, the same triple cell population distribution can be identified (Fig. 1). The mean mobility values are: 0.81, 1.20 and 1.37 μm s^{-1} V^{-1} cm^{-1}. We have demonstrated that the cells with the lowest mobility are B cells. The value 1.21 is close to that obtained for the cortisone-resistant thymic cell. According to the results we obtained using the C57Br strain, the cells with the greatest electrophoretic mobility (1.37 μm s^{-1} V^{-1} cm^{-1}) can be considered to be "educated T cells". Our experiments show that, in addition to the B cells, two types of T cells are present in the spleen of normal mice. One hypothesis would be that these two types of T cells which have the same electrophoretic mobility characteristics as those obtained for the cortisone-resistant thymic cells and for the "educated T cells" may have the same immunological functions.

It has been established that the thymus-derived cell has different antigenic and functional properties during the different stages of its thymic and post-thymic maturation.

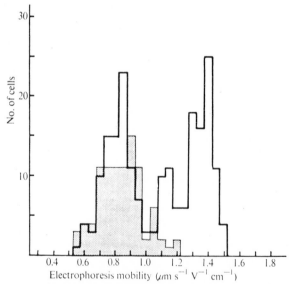

Fig. 1 Histogram of electrophoretic mobilities of homozygous and heterozygous lymphoid cells. The shaded part corresponds to the homozygous nude mice and has been superimposed on the histogram of the electrophoretic mobilities of the heterozygous nude mice spleen cells.

176

For example, the thymocytes have more 0^3 and less H2 antigen than peripheral lymphocytes[4]. In the TL positive strains, only the thymocytes express this antigen[5,17]. It has been suggested, moreover, that the thymocyte is differentiated as a T_1 cell within the thymus and then matures to a T_2 cell in the periphery[20]. Our results show that the mobility, which is determined by the cell surface nature, seems to be capable of distinguishing between B and T cells and of distinguishing certain stages in the evolution of the thymus-derived cell. On the basis of the electrophoretic mobility analysis, the following terminology is proposed: T_0, the thymocyte; T_1, the cortisone-resistant thymic cell; T_2, the "educated T cell".

These values can be used to define some functional stages in the maturation process of the thymus-derived cell, but it is possible that other steps will be found for the B and T cells in other experimental systems using electrophoretic mobility analysis.

We thank Dr J. N. Mehrishi who initiated us in cell electrophoresis techniques, Mrs Bujadoux and Miss Harter for histological controls of our cell suspensions, Mrs Heluain for the preparation of human washed erythrocytes and Mr J. McKee for correction of this text.

[1] Raff, M. C., Nature, 224, 378 (1969).
[2] Miller, J. F. A. P., and Sprent, J., Nature, 230, 267 (1971).
[3] Reif, A. E., and Allen, J. M. V., J. Exp. Med., 120, 413 (1964).
[4] Winn, H. J., Nat. Cancer Monograph, 2, 113 (1960).
[5] Aoki, T., Hammerling, U., de Harven, E., Boyse, E. A., and Old, L. J., J. Exp. Med., 130, 979 (1969).
[6] Raff, M. C., Nase, S., and Mitchison, N. A., Nature, 230, 50 (1971).
[7] Flanagan, S. P., Genet. Res., Cambridge, 8, 295 (1966).
[8] Pantelouris, E. M., Nature, 217, 370 (1968).
[9] Raff, M. C., and Wortis, H. H., Immunology, 18, 931 (1970).
[10] Janossy, G., and Greaves, M. F., Clin. Exp. Immunol., 9, 483 (1971).
[11] Blomgren, H., and Andersson, B., Cell. Immunol., 1, 545 (1971).
[12] Andersson, B., and Blomgren, H., Cell. Immunol., 1, 362 (1971).
[13] Eliott, E. V., Wallis, V., and Davies, A. J. S., Nature New Biology, 234, 77 (1971).
[14] Claman, H. N., Levine, M. A., and Cohen, J. J., Cell Interactions and Receptor Antibodies in Immune Response (edit. by Makela, O., Cross, A., and Kosunen, T. U.), 333 (Academic Press, London and New York, 1971).
[15] Mitchell, G. F., and Miller, J. F. A. P., Proc. US Nat. Acad. Sci., 59, 296 (1968).
[16] Miller, J. F. A. P., Cell Interactions and Receptor Antibodies in Immune Response (edit. by Makela, O., Cross, A., and Kosunen, T. U.), 293 (Academic Press, London and New York, 1971).
[17] Lance, E. M., Cooper, S., and Boyse, E. A., Cell. Immunol., 1, 536 (1971).
[18] Bangham, A. D., Flemens, R., Heard, D. H., and Seaman, G. V. F., Nature, 182, 642 (1958).
[19] Ruhenstroth-Bauer, G., and Lucke-Huhle, C., J. Cell. Biol., 37, 196 (1968).
[20] Raff, M. C., Transplant Rev., 6, 52 (1971).

177

LYMPHOCYTE ACTIVATION

I. RESPONSE OF T AND B LYMPHOCYTES TO PHYTOMITOGENS

G. JANOSSY AND M. F. GREAVES

SUMMARY

The selectivity of phytomitogens for T (thymus derived/dependent) and B ('bursa-equivalent' dependent/derived) lymphocytes from the mouse spleen has been investigated. Responses of normal spleen cell cultures were compared with those of cultures of selected B cells. The latter were obtained from three sources (1) spleen cells of mice that had been thymectomized as adults, lethally irradiated and reconstituted with syngeneic bone marrow cells pretreated with anti-θ serum (2) spleen cells from congenitally athymic ('nude') mice and (3) spleen cells from normal mice treated with anti-θ serum plus guinea-pig complement prior to culture.

Using a variety of different culture conditions it was shown that B cells respond well to pokeweed mitogen, and poorly if at all to phytohaemagglutinin. Responsiveness to the latter mitogen in normal spleen cell cultures appears to be a property of T cells.

INTRODUCTION

The mitogenic activity of plant lectins or phytohaemagglutinins has been extensively studied since its discovery by Nowell (1960) (reviewed by Naspitz & Richter, 1968). Although most attention has been directed towards the mitogen from *Phaseolus vulgaris* (phytohaemagglutinin or PHA), extracts from most plants of the Leguminosae family have similar properties. A considerable variety of other substances including bacterial products and antibodies activate lymphocytes in much the same fashion as phytomitogens (Ling, 1968; Oppenheim, 1968; Hirschhorn, 1970). Much of the impetus for this work derives from the observation that the gross morphological and biochemical characteristics of mitogen induced lymphocyte responses *in vitro* are very similar to antigen induced immune reactions *in vivo*. It has been suggested that the various mitogenic substances 'bypass' the requirement for antigenic recognition and induce cells to undergo that pattern of response 'normally' dependent upon strict immunological activation (Coulson & Chalmers, 1964). It is therefore generally considered that the lymphocyte activation phenomenon *in vitro* offers not only a means of analyzing the biochemical events involved in cellular de-repression (Pogo, Allfrey & Minsky, 1966; Hirschhorn *et al.*, 1969), but is also of considerable value as a clinical tool for monitoring the immunological competence of lymphocytes from both patients with various immunological disorders and those undergoing immunosuppressive therapy (Oppenheim, 1968).

It has become increasingly clear in recent years that lymphocytes are heterogeneous and that in essence a dichotomy exists between thymus-derived (T) cells and 'bursa-equivalent'-derived (B) cells (reviewed by Meuwissen, Stutman & Good, 1969; Roitt *et al.*, 1969; Davies *et al.*, 1971). It would therefore seem pertinent to inquire into selectivity, if any of various mitogens, towards T and B lymphocytes. This approach has already proved fruitful with Phytohaemagglutinin (PHA). Studies with cells from animals or human beings with selective immunological impairments, induced experimentally or occurring 'naturally', have clearly demonstrated that the responsiveness to PHA is predominantly, if not uniquely, a property of T lymphocytes (references in Table 5 in Discussion).

The phytomitogen Pokeweed (PWM) from *Phytolacca americana* has also been widely studied although to a lesser extent than PHA (Farnes *et al.*, 1964; Chessin *et al.*, 1966). One of the most interesting aspects of the response of lymphocytes to the former substance is that a considerable proportion of activated cells develop ultrastructural characteristics of the plasma cell line *in vivo* (Douglas *et al.*, 1967; Douglas & Fudenberg, 1969; Barker, Lutzner & Farnes, 1969). Since this pattern of response is a characteristic property of B lymphocytes we might therefore predict that, in contrast to PHA, PWM stimulates B cells *in vitro*. We have attempted to show that this is correct by analysing the response to PHA and PWM in mouse lymphocyte suspensions devoid of T cells. In addition to the conventional approach of thymectomy as a means of T deprivation we have taken advantage of the finding that antisera to the theta (θ) antigen in the presence of guinea-pig complement appears to be selectively toxic for T cells (Raff, 1969). Three types of T-deprived spleen cell suspensions have been studied: (1) cells from mice previously thymectomized as adults, lethally irradiated and subsequently reconstituted with syngeneic bone marrow cells which had been pretreated with anti-θ serum—referred to below as B spleen—(2) cells after direct treatment with anti-θ serum (plus complement) and (3) cells from a 'nude' strain of mouse with congenital thymic aplasia.

MATERIALS AND METHODS

Mice

Normal (control) mice. Six- to twelve-week-old inbred CBA (H-2^k) mice were used in all experiments.

T-cell deprived mice. Four-week-old CBA mice were thymectomized. Two weeks later they were lethally X-irradiated with 850 r and reconstituted i.v. 24 hr later with $5–10 \times 10^6$ viable bone marrow mononuclear cells which had been pretreated with anti-θ serum plus guinea-pig complement (under conditions which were optimal for lysing θ positive lymphocytes in spleen and lymph node). Mice were used as a source of spleen cells (B spleen cells) 3–8 weeks after reconstitution.

Congenitally athymic ('nude') mice (Flanagan, 1966; Pantelouris, 1968). These mice were

179

bred from a nucleus of mating pairs kindly provided by Dr D.S. Falconer of the Institute of Animal Genetics, Edinburgh. They were used at 4–8 weeks of age.

X-irradiation

Mice were X-irradiated with a cobalt-60 source at 68 rads/min, whole body irradiation, 190 cm from the source. They received a total of 850 r.

Lymphocyte cultures

1. *Cell sources and manipulations.* All techniques were carried on aseptically. Spleens were placed in a 90 mm Petri dish (Sterilin Ltd., Cat. No. N7) containing 5 ml cold Eagle's M.E.M. The contents of spleens were teased out with forceps. The cell suspension was pushed through a series of needles of graded sizes and transferred to a plastic tube (NUNC, Denmark). Any remaining cell aggregates were allowed to sediment for 5 min. The supernatant single cell suspension was transferred to a second tube and centrifuged 10 min at 140 g at 4°. The cells were resuspended in 5 ml medium and counted with Trypan blue. The suspension was adjusted to contain 2×10^6 living leucocytes per ml.

Some experiments involved additional treatment of cell suspensions prior to culture. To eliminate the vast majority of polymorphonuclear cells (in 'nude' mouse experiments) and macrophages (in a study of the stimulating capacity of foetal calf serum) or dead cells (after anti-θ serum treatment) the cell suspensions were filtered through cotton wool at 37°. Suspensions treated in this way contained less than 1% phagocytic cells (colloidal carbon uptake) with more than 95% viable cells.

2. *Culture media.* Two varieties of culture media have been used. (a) an Eagle's Minimal Essential Medium (M.E.M.) with double strength amino acids, supplemented with 10% heat inactivated foetal calf serum (FCS) (Rehautin, Armour Pharmaceutical Int., Kaukakee, Ill. USA), (b) RPMI-1640 medium (Flow Labs. Ltd.) supplemented with 10% heat inactivated foetal calf serum (Flow Labs. Ltd. Batch No. L40218). Both media were further supplemented with freshly prepared glutamine solution (400 mM/2 ml per 100 ml medium) and antibiotics (Penicillin 200 U/ml and Streptomycin 100 μg/ml).

3. *Cultivation conditions.* Three types of culture conditions have been used. In most of experiments the cultures were set up in screw cap Falcon tubes (12×75 mm, area of bottom approximately 120 mm^2, Gateway Int., Cat. No. 2003). 2×10^6 leucocytes were suspended in 1 ml total volume.

To determine the effect of cell density on lymphocyte stimulation a microplate method was used. The microplates (Cooke Ltd., Microtiter R) contained 96 wells with flat bottoms, each approximately 0·2 ml capacity with 26 mm^2 surface area. The examined cell density range spanned over $1 \times 10^5 - 1 \times 10^6$ leucocytes per well*. In the above two culture systems the medium consisted of 10% FCS (Rehautin) in Eagle's M.E.M.

The third type of culture system was a screw cap plastic vial with flat bottom (area 180 mm^2, 8 ml capacity, Sterilin Ltd., Cat. No. P. 118/s). 3×10^6 leucocytes were cultured in 1·5 ml RPMI-1640 with 10% FCS (Flow Labs.).

All cultures were bubbled with a gas mixture of 7% CO_2, 10% O_2 and 83% N_2, and placed into a pressure cooker containing a buffer solution. This vessel was put into thermostate of 37°. The 7% CO_2 tension was maintained during the period of cultivation.

All cultures were set up in duplicate except where stated.

* Lymphocytes were always plated and allowed to sediment before the stimulant was added.

180

4. *Measurement of responses.* In the majority of experiments both RNA and DNA synthesis were examined in the same cultures. Double labelling was performed by adding 0·017 μCi [^{14}C]uridine (57 mCi/mmol, Radiochemical Centre, Amersham) after 14 hr incubation and 0·1 μCi [^{3}H]thymidine (5Ci/mmol, Radiochemical Centre, Amersham) per 1 ml of culture fluid at 36th hr. The cultures were terminated after 60 hr. In microplate experiments [^{3}H]thymidine (0·3 μCi into each well) was added at 36 hr and the cultures harvested at the 60th hr.

After 60 hr cultivation the cell suspensions were centrifuged and the cells washed once in cold phosphate buffered saline (lacking calcium and magnesium ions). One drop 3% BSA as carrier protein was dropped onto the cell pellet and after adding 1·5 ml of cold 5% TCA the tubes were left for precipitation for at least 30 min. The precipitates were washed in 1·5 ml of cold 5% TCA followed by 2 ml cold methanol. The precipitates were washed onto membranes (GF/C 2·5 cm, Whatman) in a 'Manifold' multiple sample collector (Millipore). The membranes were left to dry and subsequently placed into scintillation vials. 0·3 ml solubilizer (Soluene TM 100, Packard) was added and after incubation in 37° overnight or at 55° for 2 hr, 5 ml scintillation fluid (5 g PPO and 0·1 g POPOP in 1 litre toluene) was put into each vial. These were kept at 4° overnight and counted in a Beckman LS-200B liquid scintillation system.

To study protein synthesis by [^{3}H]leucine uptake (L-leucine-4,5-H-3, 57 Ci/mmol, 10 μCi/ml of culture fluid, Radiochemical Centre, Amersham) the preparation of cells was essentially the same except that the use of carrier protein was omitted and after the third washing with TCA the precipitate was transferred to the membranes.

5. *Viability and cytologic examinations.* Dye exclusion tests (0·4% Trypan blue) were routinely performed on cultivated cell suspensions. Smears were regularly prepared and after fixing in methanol, stained with Giemsa.

Phytomitogens

1. *Phytohaemagglutinin (PHA).* Two different PHA preparations have been used in these experiments: (a) Wellcome PHA (PHA-W), reagent grade, lot K 2381, and (b) a purified preparation of PHA (p-PHA), kindly supplied by Dr S. Yachnin. This material gave a single band on polyacrylamide gel electrophoresis.

2. *Pokeweed mitogen (PWM).* This stimulant was purified from the plant stems (*Phytolacca americana*) using the method described by Boregson *et al.* (1966). Stimulants were stored at −20° in small aliquots.

Anti-θ serum and anti-mouse 'B' lymphocyte-antigen (MBLA) serum

These sera were a gift from Dr M. Raff. Details of their specificity and method of production here been published (Raff, 1969; Raff, Nase & Mitchison, 1971).

Anti-θ serum treatment of spleen cell suspensions

0·5 ml anti-θ serum was added to 1·5 normal spleen suspension containing 80×10^6 living leucocytes in MEM. This suspension was incubated for 30 min at 37°, washed twice in MEM, and 2 ml guinea-pig complement (1:10) was added to the cell pellet. After mixing and a further 30 min incubation in 37° the cells were washed twice and cotton filtered to remove dead cells (see above). The control cell suspension was identically treated except that normal mouse serum was used instead of anti-θ serum.

181

TABLE 1. Morphological composition of spleen cell suspensions and their survival values in tissue culture

	Normal spleen	'B' spleen	Athymic 'nude' spleen
1. Morphology (without filtration*)			
(a) Lymphocyte-like cells (%)	80	60	45–50
'B' type†	56	55	45
'T' type†	24	4–6	1–2
(b) Other cells (%)			
Large mononuclear cells	3–4	7–8	6–12
Granulocytes	10	10–20	35–40
Immature myelo, erythro- poetic and other ele- ments	5	15–25	4–5
2. Lymphocyte survival‡ in culture (%)			
0 hr	100	100	
24 hr	77(90§)	61	n.t.
48 hr	45	35	
60 hr	30(50§)	26	

* After cotton wool filtration (see Materials and Methods) all suspensions contained 90–95% lymphocytes.

† Calculated from the % theta (θ) positive and MBLA positive cells.

‡ Tube cultures, Eagle's medium (expressed as a proportion of viable cell number at 0 hr, at the start of cultures the suspensions contained 90% viable cells).

§ Flat bottomed culture vessels. RPMI-1640 medium (see Materials and Methods). Both 3 and 4 assayed by Trypan blue dye exclusion.

n.t. = not tested.

RESULTS

Characterization of spleen cell suspensions

The three types of spleen cell suspensions used differed in their content of various cell types (Table 1). The approximate proportions of 'T' and 'B' type lymphocytes was calculated from the percentage cytotoxicity induced by anti-θ (i.e. anti-T cell) and anti-MBLA (i.e. anti-B cell) sera in the presence of guinea-pig complement. Although the sum of the proportions of cells killed by these two sera was always approximately equal to the *total* proportion of lymphoid cells it is likely that a sizable proportion of 'lymphocyte-like' cells in B spleens are in fact precursors of haemopoetic cells.

As shown in Table 1 B spleens and athymic ('nude') spleens had a greater proportion of non-lymphoid cells than control normal spleens.

Survival values for the different cell suspensions in culture are also given in Table 1. The survival of B spleen lymphocytes over a 60 hr period was only marginally inferior to that of cells from normal spleens (cf. Doenhoff *et al.*, 1970). Cell viability in flat bottomed vial cultures (with RPMI medium) was consistently better than in tube cultures (with

182

Eagle's medium) (cf. Adler *et al.*, 1970). No attempt was made to monitor selective survival rates of T and B lymphocytes in the normal spleen cell cultures.

Responses of T and B lymphocytes to phytohaemagglutinin (PHA) and pokeweed mitogen (PWM)

1. *'B' spleen cells compared with normal control spleen cells.* The response of different spleen cell suspensions to varying concentrations of p-PHA and PWM was studied using the tube culture technique (see Materials and Methods). Both RNA synthesis ($[^{14}C]$uridine uptake) and DNA synthesis ($[^3H]$thymidine uptake) were quantitated. The results are

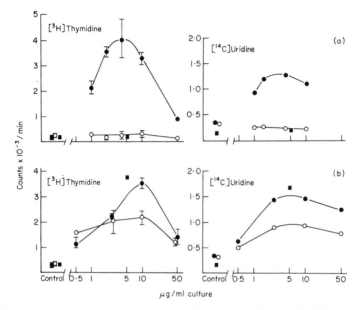

FIG. 1. Activation of lymphocytes by phytomitogens. (a) purified phytohaemagglutinin (p-PHA), (b) pokeweed mitogen (PWM). ●, normal spleen cells; ○, 'B' spleen cells; ■, athymic ('nude') spleen cells. Vertical bars represent 1 standard error of mean. Mean values of two to five cultures given.

shown in Fig. 1. PHA stimulated both RNA and DNA synthesis in the normal spleen cell cultures but induced no significant response in B spleen cell cultures. In contrast, PWM activated both normal and B cultures, the former giving the greater response to the optimal concentration of mitogen*. High concentrations of mitogens induced sub-optimal responses. Viability studies on cultures treated with high mitogen concentrations suggested that reduced responsiveness was not attributable to overt toxicity (PHA-W, in contrast to p-PHA, was however toxic to cells at these concentrations). In further experiments PHA concentrations

* It should, however, be noted that absolute comparison of responsiveness is not valid since the proportion of lymphocytes and contaminating granulocytes in the two groups was unequal.

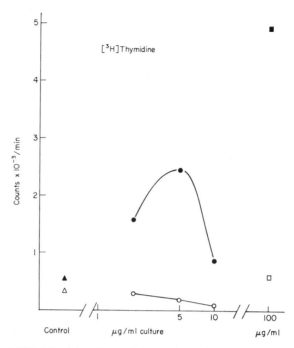

FIG. 2. Early PHA-induced thymidine uptake by lymphocytes (24–28 hr). Controls: ▲, normal spleen cells; △, 'B' spleen cells. p-PHA: ●, normal spleen cells; ○, 'B' spleen cells. PHA-W ■, normal spleen cells; □, 'B' spleen cells (the concentration of PHA-W used was optimal for thymidine uptake: 100 µg/ml, the dose/response curve for PHA-W was similar to that previously reported by Tursi *et al.*, 1969).

TABLE 2. Early response of spleen cell cultures to phytohaemagglutinin

	Response, isotope uptake c/min*			
	Normal spleen cells†		'B' spleen cells†	
Mitogen	[³H]leucine	[¹⁴C]uridine	[³H]leucine	[¹⁴C]uridine
Control	926± 61	604±48	772±49	214±17
PHA-W‡	1950±105	2159±67	898±26	334±51

* Mean±standard error of mean. Isotope added at 0 hr, and cultures harvested after 24 hr.
† Spleen cells unfiltered. Tube cultures.
‡ At 100 µg/ml (previously determined optimal concentration for [³H]thymidine uptake, 36–60 hr).

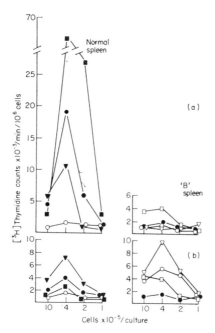

FIG. 3. Effect of cell concentration on response to phytomitogens. (a) phytohaemagglutinin, normal spleen: ■, 100 μg/ml PHA-W; ●, 20 μg/ml p-PHA; ▼, 4 μg/ml p-PHA; ○, controls, 'B' spleen; □, 100 μg/ml PHA-W; ○, 20 μg/ml p-PHA; ▽, 4 μg/ml p-PHA; ●, controls. (b) pokeweed mitogen, normal spleen: ■, 50 μg/ml PWM; ●, 10 μg/ml PWM; ▼, 2·5 μg/ml PWM; ○, controls. 'B' spleen: □, 50 μg/ml PWM; ○, 10 μg/ml PWM; ▽, 2·5 μg/ml PWM; ●, controls.

as low as 0·005 μg p-PHA per ml also failed to activate B cells which are therefore, under the described conditions, resistant to activation by a 10,000-fold range of mitogen concentrations. We next investigated whether varying any of the culture conditions would permit a B cell response to PHA to be expressed. The first point we considered was whether an 'early' (0–48 hr) response of B cells had been missed owing to the design of the experiment. To exclude this possibility we studied early uridine, thymidine and leucine (into proteins) uptake in PHA treated normal and B spleen cultures. The results are given in Table 2 and Fig. 2. These show that during the 48 hr of culture with PHA only very marginal net RNA, DNA or protein synthesis occurred in B spleen cultures in contrast to normal spleen cell cultures which showed uptake values considerably above background.

Although normal spleen and B spleen cultures had similar viability after 60 hr incubation it is theoretically possible that T and B cells have different requirements for optimal *responsiveness in vitro* to proliferative stimuli. As an empirical approach to this problem we have studied the effect of varying cell density and culture media. The effect of different lympho-

185

TABLE 3. The effect of filtration on incorporation of [³H]thymidine by normal and 'B' spleen cell cultures

Mitogen†	Expt	Response, isotope uptake c/min*			
		Normal spleen cells		'B' spleen cells	
		non-filtered	filtered	non-filtered	filtered
Control	1	1630	766	1820	480
	2	1000	510	1605	382
	3	1790	520	1230	700
PHA-W	1	35200	34300	2860	530
	2	19800	24100	3300	670
	3	35100	39700	5780	1180
PWM	1	24200	21080	4830	5370
	2	12600	10200	6630	6010
	3	28220	16800	11900	6900

* [³H]thymidine added between 36–60 hr of culture. Cultures set up in RPMI-1640/Flow foetal calf serum media in 'Sterilin' flat bottom vials (see Materials and Methods).
† Added at previously determined optimal concentrations (PHA-W: 100 μg/ml; PWM: 5 μg/ml).

cyte concentrations* on PHA and PWM induced thymidine uptake (36–60 hr) is shown in Fig. 3. In this microplate experiment (see Materials and Methods) responsiveness of normal spleen cells to PHA was critically dependent upon cell density as previously reported by Moorhead et al. (1967) 4×10^5 cells per 0·2 ml culture fluid appeared to be optimal for both mitogens. The cruder PHA-W preparation induced a considerably better response than p-PHA. The former contains erythroagglutinating activity and this may well have been in part responsible for its greater potency since Yachnin et al. (1971) have shown that coating red cells with PHA enhances mitogenicity of the latter.

High concentrations of B spleen cells gave a very small response (approximately 6% of normal spleen) to PHA-W but no response to p-PHA.

Responses to PWM were also dependent upon cell concentration. Normal and B spleen cultures gave essentially similar responses over the varying cell densities and mitogen concentrations. Although only a ten-fold range of cell densities has been studied these results imply that this factor is unlikely to be limiting the PHA responsiveness of B cells.

Earlier experiments had shown that cell survival and background thymidine uptake values were greater when cultures were set up in flat bottomed vials with RPMI-medium as compared to the standard tube culture method with Eagle's medium†. The response of normal and B spleen cells to PHA-W and PWM cultured under these improved conditions is shown in Table 3. In addition the effect of cotton filtration has been studied. It can be

* The relevant parameter being varied is probably cell density per unit area of the culture vessel bottom (cf. Moorhead, Connolly & McFarland, 1967; Watkins & Moorhead, 1969).
† Note that these two media were supplemented with different sources of foetal calf serum (see Materials and Methods).

TABLE 4. Response of normal and athymic ('nude') spleen to phytomitogens *in vitro*

	Response, isotope uptake c/min*			
	Normal spleen cells		'B' spleen cells	
Mitogen†	non-filtered	filtered	non-filtered	filtered
Control	1730	800	540	230
p-PHA	4230	6400	230	280
PHA-W	7600	n.t.	180	n.t.
PWM	15300	10130	1900	7890

* Each value is the mean of duplicate tube cultures (variation between duplicates was < 20%). [³H]thymidine, 36–60 hr of culture.

† Added at previously determined optimal proportions (p-PHA: 5 μg/ml; PHA-W: 100 μg/ml; PWM: 5 μg/ml).

n.t. = not tested.

seen that non-filtered B spleen cultures did in fact give a small response to PHA which was approximately twice background. In one case the PHA induced response of B spleen was more marked. Cotton filtration (see Materials and Methods) of the cells prior to culture had three interesting effects: (a) background uptake was considerably reduced, (b) the PHA induced response of B spleen, but not of normal spleen cultures, was lost and (c) PWM induced responsiveness of B spleen cultures was unimpaired. This experiment suggests, therefore, that under certain culture conditions a small PHA induced response of B spleen cells can be shown which is apparently dependent to a large extent upon the presence of 'adhesive' cells. There are several possible interpretations of this observation (see Discussion). It is clear however that under the different conditions of culture investigated the response of B spleen cells to PHA was always less than 15% of the normal spleen cultures, despite the fact that both responded well to PWM.

2. *Responses of athymic 'nude' spleen cells compared with normal spleen cells.* Spleens of congenitally athymic 'nude' mice have been shown to be deficient in T cells (Raff & Wortis, 1970). The majority of lymphoid cells in 'nude' spleens have the surface characteristics of B cells, i.e. θ-negative, MBLA positive (see Table 1 and Raff & Wortis, 1970). Furthermore, Wortis (1971) has shown that, in contrast to normal (or heterozygous) litter mate controls, 'nude' mice do not produce a lymph node proliferative response following the injection of PHA.

The responsiveness of 'nude' spleen cells to p-PHA and PWM *in vitro* is shown in Table 4. In contrast to normal spleen cell cultures, spleen cells from 'nude' mice were not activated by PHA. They did however respond to PWM. Reactivity was considerably enhanced by cotton filtration prior to culture implying that granulocytes or other adhesive cells present in considerable numbers in spleens of these athymic mice (see Table 1) may depress the proliferative response of B-cells.

3. *Affect of anti-θ (theta) serum on the responsiveness of normal spleen cell suspensions to PHA and PWM.* Anti-θ serum selectively lyses T cells in the presence of guinea-pig complement (Raff, 1969). We therefore anticipated that the treatment of normal spleen cells

187

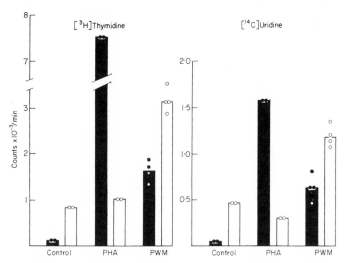

FIG. 4. Effect of anti-θ serum on the response of spleen cells to phytomitogens. ■, cells pretreated with control (AKR) serum plus complement, □, cells pretreated with anti-θ serum plus complement. Each point represents an individual culture.

with these reagents prior to culture would abolish the PHA response. As shown in Fig. 4 the PHA response was in fact reduced by 90% by anti-θ serum. In marked contrast, the PWM response was enhanced by pretreatment with anti-θ serum implying that T cells may make little contribution to the PWM response of normal spleen cell cultures. Thymidine and uridine uptake were similarly affected. It is also interesting to note that anti-θ serum plus complement itself induced a proliferative response (approximately five times background). We suppose that this effect is due to activation of that small proportion of T cells which escaped the lytic effect of anti-θ serum. It will be of interest in this respect to study the activation of lymphocytes by anti-θ serum under non-lytic conditions.

DISCUSSION

These results demonstrate that whereas phytohaemagglutinin (PHA) selectively activates thymus-derived (T) cells, pokeweed mitogen (PWM) activates 'bursa-equivalent'-derived (B) cells. T cells may also respond to PWM although the evidence for this in the above experiments was equivocal*. The selective nature of activation induced by the two mitogens was manifested by RNA, DNA and protein synthesis, and was evident over several days in tissue culture with a wide range of mitogen concentrations. The results were essentially the same with all three sources of B cell suspensions: (a) anti-θ serum treated 'normal' spleens, (b) congenitally athymic mouse spleens and (c) spleens from adult thymectomized mice, lethally irradiated and reconstituted with anti-θ serum treated syngeneic bone marrow suspensions (B-spleens).

* Recent experiments (Janossy & Greaves, unpublished) with thymocytes from cortisone-treated mice have shown that PWM does in fact stimulate T cells. *Clin. exp. Immunol.* (1972) **10**, 525–536.

TABLE 5. A summary of direct and circumstantial evidence for selective activation of 'T' and 'B' lymphocytes by phytohaemagglutinin and pokeweed mitogen

Test system	Species	PHA	PWM	References
1. Cell markers				
i. chromosomal				
T6T6	mouse	T		Doenhoff *et al.* (1970)
sex (y)	rat	T		Johnson & Wilson (1970)
ii. Theta (θ) surface antigen—susceptibility to anti-θ serum				
(a) Pretreatment of cells prior to culture	mouse	T	B	This study*
(b) Cytotoxicity of stimulated cells	mouse	T		Owen, J.J.T. (personal communication)
2. Experimental immunological deprivation				
i. neonatal thymectomy	mouse	T		Martial-Lasforgues *et al.* (1966), Dukor & Dietrich (1967), Rodey & Good (1970), Takiguchi *et al.* (1971)
	rat	T		Rieke (1966), Schwarz (1968), Meuwissen *et al.* (1969b)
	chicken	T		Greaves, Roitt & Rose (1968), Meuwissen *et al.* (1969a)
ii. Adult thymectomy plus anti-lymphocyte globulin	mouse	T		Tursi *et al.* (1969)
iii. Adult thymectomy plus 'B' cell reconstitution	mouse	T	B	This study
iv. Bursectomy	chicken	T		Greaves *et al.* (1968), Alm & Peterson (1969), Meuwissen *et al.* (1969a)
3. Congenital or 'naturally' acquired immunological defiiency				
i. cell-mediated e.g. thymic aplasia	mouse	T	B	This study
	man	T		Oppenheim (1968), Gotoff (1968), Douglas, Kamin & Fudenberg (1969), Bach *et al.* (1969)
ii. humoral e.g. agammaglobulinaemia (Bruton)	man	T	B	
4. Tissue distribution of responsive cells				
i. thymus cf. bursa	chicken	T		Weber (1967)
ii. thymus cf. peripheral tissue/circulating cells	rat	T		Colley, Shih Wu & Waksman (1970)
	man	T		Claman (1966), Pegrum, Ready & Thompson (1968)
iii. Gut associated ('B' type ?) tissue cf. other lymphoid tissue	mouse	T		Rodey & Good (1970)
5. Characteristics of the response				
i. Polyribosomal vs. polydisperse RNA synthesis	man	T	B	Chessin *et al.* (1966)
ii. Ultrastructure—rough endoplasmic reticulum	man	T	B	Chessin *et al.* (1966), Barker, Lutzer & Farnes (1969)
iii. Immunoglobulin synthesis	mouse	T	B	Parkhouse, Janossy, & Greaves (to be published)
	man	T	B	Greaves & Roitt (1968)

* In a recent study, Blomgren & Svedmyr (1971) have also described the inhibitory effect of anti-θ serum in the PHA response.

189

Unfiltered B-spleen suspensions were heterogeneous and contained a high number of myelo- and haemopoetic elements and other cells. However morphological studies on cultured cells (Janossy & Greaves, unpublished) and an analysis of immunoglobulin synthesis (Parkhouse, Janossy & Greaves, unpublished) suggest that most, if not all, of the PWM induced isotope uptake is due to B lymphocyte activation and concomitant differentiation along the plasma cell line (cf. Douglas *et al.*, 1967). In addition cotton wool filtered suspensions containing 90–99% lymphocytes responded to PWM. Despite the presence of a small proportion of residual T (θ positive) cells in these suspensions (1–6%), isotope uptake was not enhanced by PHA under the standard tube culture conditions.

The observed specificity of PHA for T cells is in accord with other direct and circumstantial evidence which has been tabulated in Table 5. In the present experiments B cells were cultured in the absence of T cells and the results do not rule out the possibility that B cells will respond either directly or indirectly when mixed with T cells. Other studies, however, using chromosomal and antigenic markers (Table 5) suggest that B cells make, at the most, only a very limited mitotic response to PHA even in the presence of activated T cells.

Previous evidence (Table 5) for responsiveness of B cells to PWM has been equivocal, although Douglas *et al.* (1969) described a depressed response of lymphocytes from agammaglobulinaemic individuals which nevertheless reacted to PHA. The experiments described above showed that purified B cell cultures do respond well to PWM. The lack of responsiveness of B cells to PHA cannot therefore be attributed to poor viability or lack of proliferative capacity *in vitro*. We feel that these observations provide an explanation for some of the observed discrepancies in the characteristics of the response of cells to PHA and PWM (Table 5). In particular it provides a basis for the observation that PWM stimulated cells synthesize polyribosomal RNA and develop rough endoplasmic reticulum (Chessin *et al.*, 1966; Douglas *et al.*, 1967)—features characteristic of the plasma cell (B) line. Earlier studies with human lymphocytes suggested that PWM but not PHA activated immunoglobulin synthesis in a small proportion of the cultured cells (Greaves & Roitt, 1968). We are currently investigating immunoglobulin synthesis and secretion in mouse lymphocyte cultures in the presence of PHA and PWM and preliminary results confirm the above mentioned study in that PWM cultures synthesize and secrete considerably more immunoglobulin (principally IgM) than controls whereas little or no enhanced synthesis was observed with PHA stimulated cultures.

It seems reasonable to conclude that PWM activated B cells, or at least a substantial proportion of them, undergo a similar pattern of differentiation as they would normally do only when stimulated by ligands functioning as specific antigens. Similarly the pattern of T cell responsiveness to PHA is similar, if not identical, to that observed when the same population is actively engaged in cellular hypersensitivity reactions *in vivo* or *in vitro*, induced by antigen.

While this study has shown that selective effects of phytomitogens exist, we do not wish to conclude that specificity of the stimulants is necessarily absolute. Although we have investigated such parameters as cell density, culture media, culture tubes/containers etc. in addition to time course studies and the effects of a wide range of mitogen concentrations (10^4), it is still theoretically possible that conditions might be found under which both populations would respond to either phytomitogen. In one of the three culture methods we investigated a very small but significant response to PHA was observed with B spleen cell suspensions. This method involved a foetal calf serum with a high stimulating capacity as manifested by the high background control values for thymidine uptake. Cell survival and responsiveness was also facilitated by the use of a flat bottomed culture vessel and RPMI-1460 medium. Background values and the PHA induced increase of thymidine uptake were both considerably reduced by cotton wool filtration of the suspensions prior to culture. We suspect that the high background and slight PHA response of unfiltered cells is attributable to a macrophage dependent immune response to foetal calf serum proteins and perhaps also to PHA itself. Despite the lack of PHA enhanced isotope uptake in cotton wool filtered B spleen cell suspensions, we have observed some cellular transformations involving 7–10% of the cells after 60 hr culture. The morphological appearance of these cells suggests that they are plasmablasts and they are therefore probably indicative of a small abortive PHA response of a minor population of B cells rather than a reaction of the small residual T cell population.

The biochemical basis of the selectivity of phytomitogens is unresolved. In this respect it is of interest to quantitate binding sites for PHA and PWM on T and B cells since these could provide a basis for the observed specificity. However our preliminary results suggest that the unresponsiveness of B cells to PHA cannot be ascribed to a lack of PHA binding sites. Another possibility is that the strong agglutination of B cells may have inhibited their response to PHA; PWM is a much weaker leucoagglutinin.

Caution is necessary in extrapolating these results to other species, although we suspect the observed relationship of PHA/T cells and PWM/B cells to be true for lymphocytes for human beings (cf. Douglas *et al.*, 1969). The rat in particular may be different in that the PWM response is almost completely ablated by thymectomy (Schwarz, 1968; Meuwissen *et al.*, 1969).

We suggest that studies on selective activation of T and B cells by phytomitogens may provide clues as to the nature of antigen induced activation and differentiation of these cell lines *in vivo*. In addition responses to PHA, PWM and other stimulants may serve as useful indicators of the proliferative potential of T and B cells in diseases or clinical procedures associated with immunological deficiency.

ACKNOWLEDGMENTS

This work was supported by a Royal Society Scholarship to G.J. and the Medical Research Council. We are grateful to Dr M. Raff for supply of antisera and Dr S. Yachnin for purified phytohaemagglutinin. We also thank Miss P. Haria and Miss F. Rose for excellent technical assistance.

REFERENCES

ADLER, W.H., TAKIGUCHI, T., MARSH, B. & SMITH, R.T. (1970) Cellular recognition *in vitro*. I. Definition of a new technique and results of stimulation by PHA and specific antigens. *J. exp. Med.* **131**, 1049.

ALM, G.V. & PETERSON, R.D.A. (1969) Antibody and immunoglobulin production at the cellular level in bursectomised irradiated mice. *J. exp. Med.* **129**, 1247.

BACH, F.H., MEUWISSEN, H.J., ALBERTINI, R.J. & GOOD, R.A. (1969) Agammaglobulinaemic leukocytes—their *in vitro* reactivity. *Proceedings of the 3rd Annual Leucocyte Culture Conference* (Ed. by W.O. Rieke), p. 709. Appleton–Century–Crofts. New York.

BARKER, B.E., LUTZNER, M.A. & FARNES, P. (1969) Ultrastructural properties of pokeweed stimulated leukocytes *in vivo* and *in vitro*. *Proceedings of the 3rd Annual Leucocyte Culture Conference* (Ed. by W.O. Rieke), p. 588. Appleton–Century–Crofts. New York.

BLOMGREN, H. & SVEDMYR, E. (1971) Evidence for thymic dependence of PHA-reactive cells in spleen and lymph nodes and independence in bone marrow. *J. Immunol.* **106**, 835.

BORJESON, J., REISFIELD, R., CHESSIN, L.N., WELSH, P. & DOUGLAS, S.D. (1966) Studies on human peripheral blood lymphocytes *in vitro*. I. Biological and physicochemical properties of the pokeweed mitogen. *J. exp. Med.* **124**, 859.

CHESSIN, L.N., BORJESON, J., WELSH, P.D., DOUGLAS, S.T. & COOPER, H.L. (1966). Studies on human peripheral blood lymphocytes *in vitro*. II. Morphological and biochemical studies on the transformation of lymphocytes by pokeweed mitogen. *J. exp. Med.* **124**, 873.

CLAMAN, H.N. (1966) Human thymus cell cultures—evidence for two functional population. *Proc. Soc. exp. Biol. (N.Y.)*, **121**, 236.

COLLEY, D.G., SHIH WU, A.Y. & WAKSMAN, B.H. (1970) Cellular differentiation in the thymus. III. Surface properties of rat thymus and lymph node cells separated on density gradients. *J. Exp. Med.* **132**, 1107.

COULSON, A.S. & CHALMERS, D.G. (1964) Effects of phytohaemagglutinin on leucocytes. *Lancet* **ii**, 819.

DAVIES, A.J.S., LEUCHARS, E., WALLIS, V. & DOENHOFF, M.J. (1971) A system for lymphocytes in the mouse. *Proc. roy. Soc.* B. **176**, 369.

DOENHOFF, M.J., DAVIES, A.J.S., LEUCHARS, E. & WALLIS, W. (1970) The thymus and circulating lymphocytes of mice. *Proc. roy. Soc.* B. **176**, 69.

DOUGLAS, S.D. & FUDENBERG, H.H. (1969) The *in vitro* development of plasma cells from lymphocytes following Pokeweed mitogen stimulation: a fine structural study. *Exp. Cell Res.* **54**, 277.

DOUGLAS, S.D., HOFFMAN, P.F., BORJESON, J. & CHESSIN, L.N. (1967) Studies on human peripheral blood lymphocytes *in vitro*. III. Fine structural features of lymphocyte transformation by Pokeweed mitogen. *J. Immunol.* **98**, 17.

DOUGLAS, S.D., KAMIN, R.M. & FUDENBERG, H.H. (1969) Human lymphocyte response to phytomitogens *in vitro*: normal, agammaglobulinaemic and paraproteinaemic individuals. *J. Immunol.* **103**, 1186.

DUKOR, P. & DIETRICH, F.M. (1967) Impairment of phytohaemagglutinin induced blastic transformation in lymph nodes from thymectomized mice. *Int. Arch. Allergy*, **32**, 521.

FARNES, P., BARKER, B.E., BROWNHILL, L.E. & FANGER, H. (1964) Mitogenic activity in *Phytolacca americana* (pokeweed). *Lancet*, **ii**, 1100.

FLANAGAN, S.P. (1966) Nude, a new hairless gene with pleiotrophic effects in the mouse. *Genet. Res.* **8**, 295.

GOTOFF, S.P. (1968) Lymphocytes in congenital immunological deficiency diseases. *Clin. exp. Immunol.* **3**, 843.

GREAVES, M.F. & ROITT, I.M. (1968) The effect of phytohaemagglutinin and other lymphocyte mitogens on immunoglobulin synthesis by human peripheral blood lymphocytes *in vitro*. *Clin. exp. Immunol.* **3**, 393.

GREAVES, M.F., ROITT, I.M. & ROSE, M.E. (1968) Effect of bursectomy and thymectomy on the response of chicken peripheral blood lymphocytes to phytohaemagglutinin. *Nature (Lond.)*, **220**, 293.

HIRSCHHORN, K. (1930) Situations, leading to lymphocyte activation. *Mediators of Cellular Immunity* (Ed. by H.S. Lawrence and M. Landy), p. 1. Academic Press, New York.

HIRSCHHORN, R., TROLL, W., BRITTINGER, G. & WIESSMANN, G. (1969) Template activity of nuclei from stimulated lymphocytes. *Nature (Lond.)*, **222**, 1247.

JOHNSTON, J.M. & WILSON, D.B. (1970) Origin of immunoreactive lymphocytes in rats. *Cell. Immunol.* **1**, 430.

LING, N.R. (1968) *Lymphocyte stimulation*. North Holland Publ. Co. Amsterdam.

MARTIAL-LASFARGUES, C., LIACOPOLOUS-BRIOT, M. & HALPERN, B.M. (1966) Culture des leucocytes sanguins de souris *in vitro*. Etude de l'action de la phytohémagglutinine sur les lymphocytes de souris normales et thymectomisées. *C.R. Soc. Biol.* **160**, 2013.

MEUWISSEN, H.J., STUTMAN, O. & GOOD, R.A. (1969) Functions of the lymphocytes. *Seminars in Haematol.* **6**, 28.

MEUWISSEN, H.J., VAN ALTEN, P.A., COOPER, M.D. & GOOD, R.A. (1969a) Dissociation of thymus and bursa function in the chicken by PHA. *Proceedings of the 3rd Annual Leucocyte Culture Conference* (Ed. by W.O. Rieke), p. 227. Appleton–Century–Crofts. New York.

MEUWISSEN, H.J., VAN ALTEN, P.A. & GOOD, R.A. (1969b) Decreased lymphoid cell multiplication in the post-thymectomy state. *Transplantation*, **7**, 1.

MOORHEAD, J.F., CONNOLLY, J.J. & McFARLAND, W. (1967) Factors affecting the reactivity of human lymphocytes *in vitro* I. Cell number, duration of culture and surface area. *J. Immunol.* **99**, 413.

NASPITZ, C.K. & RICHTER, M. (1968) The action of Phytohaemagglutinin *in vivo* and *in vitro*—a review. *Progr. Allergy*, **12**, 1.

NOWELL, P.C. (1968) Phytohaemagglutinin: an initiator of mitosis in cultures of human leukocytes. *Cancer Res.* **20,** 462.

OPPENHEIM, J.J. (1968) Relationship of *in vitro* transformation to delayed hypersensitivity in guinea pigs and man. *Fed. Proc.* **27,** 21.

PANTELOURIS, E.M. (1968) Absence of the thymus in a mouse mutant. *Nature (Lond.),* **217,** 370.

PEGRUM, G.D., EADY, D.R. & THOMSON, E. (1968) The effect of PHA on human foetal cells grown in culture. *Brit. J. Haemat.* **15,** 371.

POGO, B.G.T., ALLFREY, V.G. & MIRSKY, A.E. (1966) RNA synthesis and histone acetilation during the course of gene activation in lymphocytes. *Proc. nat. Acad. Sci. (Wash.),* **55,** 805.

RAFF, M.C. (1969) Theta isoantigen as a marker of thymus derived lymphocytes in mice. *Nature (Lond.),* **224,** 378.

RAFF, M.C., NAZE, S. & MITCHISON, N.A. (1971) Mouse specific 'B' lymphocyte antigen (MBLA)—a marker for thymus independent lymphocytes. *Nature (Lond.),* (in press).

RAFF, M.C. & WORTIS, H.H. (1970) Thymus dependence of θ-bearing cells in the peripheral tissues of mice. *Immunology,* **18,** 931.

RIEKE, W.O. (1966) Lymphocytes from thymectomized rats: immunologic, proliferative and metabolic properties. *Science,* **152,** 535.

RODEY, G.E. & GOOD, R.A. (1970) The *in vitro* response to PHA of lymphoid cells from normal and neonatally thymectomized adult mice. *Int. Arch. Allergy,* **36,** 399.

ROITT, I.M., GREAVES, M.F., TORRIGIANI, G., BROSTOFF, J. & PLAYFAIR, J.H.L. (1969) The cellular basis of immunological responses. *Lancet,* **ii,** 367.

SCHWARZ, M.R. (1968) Transformation of rat small lymphocytes with pokeweed mitogen (PWM). *Anat. Rec.* **160,** 47.

TAKIGUCHI, T., ADLER, W.H. & SMITH, R.T. (1971) Cellular recognition *in vitro* by mouse lymphocytes. Effects of neonatal thymectomy and thymus graft restoration on alloantigen and PHA stimulation of whole and gradient separated sub-population of spleen cells. *J. exp. Med.* **133,** 63.

TURSI, A., GREAVES, M.F., TORRIGIANI, G., PLAYFAIR, J.H.L. & ROITT, I.M. (1969) Response to phytohaemagglutinin of lymphocytes from mice treated with anti-lymphocyte globulin. *Immunology,* **17,** 801.

WATKINS, S.M. & MOORHEAD, J.F. (1969) The effect of cell crowding on the *in vitro* reactivity of normal and abnormal human lymphocytes. *Cell Tissue Kinet.* **2,** 213.

WEBER, W.T. (1967) The response to Phytohaemagglutinin by lymphocytes from the spleen, thymus and bursa of Fabricius of chicken. *Exp. Cell Res.* **46,** 464.

WORTIS, H.H. (1971) Immunological responses of 'nude' mice. *Clin. exp. Immunol.* **8,** 305.

YACHNIN, S., ALLEN, L.W., BARON, J.M. & SVENSON, R. (1971) Potentiation of lymphocyte transformation by membrane–membrane interaction. *Proceedings of the 4th Annual Leucocyte Culture Conference* (Ed. by O.R. McIntyre), p. 37. Appleton–Century–Crofts. New York.

Activation of T and B Lymphocytes by Insoluble Phytomitogens

MELVYN F. GREAVES

& SARA BAUMINGER

Although insoluble phytomitogens are probably not taken within cells, they are shown here to activate both T and B cells. It is conceivable therefore that antigens also stimulate lymphocytes through a similar membrane event.

THE initial and critical event in lymphocyte stimulation by both specific (antigen) and non-specific (for example, phytomitogen and antibody) ligands is the combination of the latter with binding sites on or in the lymphocyte. There are theoretical reasons for supposing that these receptors should be exposed on the plasma membrane and indeed the evidence for this with respect to both categories of stimulants is now compelling[1-5]. However, there is also evidence to suggest that subsequent to binding at the cell surface, phytohaemagglutinin (PHA)[6], anti-immunoglobulin antibodies[7] and antigen[8] may in fact be internalized, probably by pinocytosis. This raises important questions concerning the precise site and mode of action of the stimulating molecules. It has been suggested, for example, that the rapid nuclear protein changes (such as histone acetylation[9] or exposure of phosphate groups[10]) induced by PHA are related to the intranuclear chromatin localization of this mitogen in lymphocytes[11]. An extranuclear localization of ligand receptors does not preclude an intranuclear site of primary action—as demonstrated in the instance of oestrogen hormone responsive cells[12].

We have investigated the capacity of insoluble and presumably non-internalizable phytomitogens to stimulate lymphocytes. PHA and pokeweed mitogen (PWM) have been covalently linked to 'Sepharose' beads and added to mouse lymphocytes from different tissues. The responsiveness of purified T and B lymphocytes in terms of radioactive uridine and thymidine uptake has been compared with that induced by the same mitogens in a soluble form.

Responsiveness of T and B Cells to PHA and PWM-'Sepharose'

The effect of varying concentrations of soluble and 'Sepharose'-bound mitogens on RNA and DNA synthesis by T, B and unselected (T plus B) cell suspensions is shown in Table 1. The PWM-'Sepharose' beads were slightly more effective in activating unselected (T plus B) and B cells than the soluble compound; T cells, however, responded better to the soluble PWM. Three different batches of PHA-'Sepharose' were tested, and although absolute responses varied all behaved dramatically different from soluble PHA. B cells were completely unresponsive to soluble PHA but were clearly activated by PHA on 'Sepharose'. This effect is unlikely to be due to a precocious response by the small residual T cell population in these suspensions (1–5%), since pretreatment of cultures with anti-θ (C3H) serum plus guinea-pig complement to eliminate T cells had no effect. A similar treatment of unselected spleen cells eliminated the response to soluble PHA, but only marginally reduced the response to 'Sepharose'-PHA (Table 2), indicating that a considerable part of the unselected spleen cell response to 'Sepharose'-PHA was by B cells. 'Sepharose' beads remained intact during the culture period and although initially many lymphocytes adhered to the mitogen-coated bead surfaces they later detached and the majority of enlarged "blast" cells formed were free in the medium. BSA-'Sepharose' induced only very marginal or negligible responses, thus tending to exclude the possibility that the phytomitogens on 'Sepharose' were functioning as "super-antigens"[20].

Controls

A major difficulty in experiments with insoluble stimulants is to demonstrate unequivocally that the effects observed were not due to soluble derivatives leaking off the insoluble matrix. Although PHA and PWM were coupled to 'Sepharose' through covalent linkages and were thoroughly washed before use, there are at least two potential ways in which bound molecules could be released during culture. First, enzymatic (galactosidase, protease) activity of lymphocytes might release mitogen

195

which could then be taken up directly by cells or through the culture medium. Second, phytomitogens have specificity for saccharides[21] and it is therefore possible that some of the initial binding to 'Sepharose' might have been through non-covalent reaction with galactoside residues. On subsequent incubation such non-covalently bound molecules might be slowly released and competitively taken up by lymphocytes. There is evidence to suggest, however, that neither of these factors was significant in this study. More than 98% of trace-labelled mitogen remained bound to the 'Sepharose' during a 48 h incubation period at $37°$ C, with or without lymphocytes. The maximal amount of active soluble mitogen released in these experiments is insufficient to account for the observed response and indeed supernatants from 'Sepharose'-PHA beads incubated for 48 h at $37°$ C were found to have effectively no mitogenic activity. We have also added PHA to non-activated 'Sepharose'. The beads were then treated as described for the covalently linked

Table 1 Activation of T and B Lymphocytes by Soluble and Insoluble Phytomitogens

Preparation	Response: maximum uptake (counts $\times 10^{-3}$/min) *		
	Unselected (T + B) cells	T cells	B cells
Controls (untreated)	896 ± 260	856 ± 210	963 ± 270
Soluble PHA	$24,514 \pm 2,620$	$28,456 \pm 3,260$	$1,140 \pm 461$
'Sepharose'-PHA (1)	$35,430 \pm 2,310$	NT	$30,154 \pm 2,100$
(2)	$20,646 \pm 2,160$	$6,926 \pm 2,100$	$3,834 \pm 1,050$
(3)	$41,695 \pm 4,280$	$19,350 \pm 1,080$	$39,250 \pm 3,140$
'Sepharose'-PHA (3) supernatant ‡	$1,624 \pm 160$	NT	NT
Non-Co-V 'Sepharose'-PHA †	$1,948 \pm 426$	NT	NT
Soluble PWM	$24,982 \pm 2,780$	$12,756 \pm 1,550$	$15,646 \pm 2,940$
'Sepharose'-PWM	$31,861 \pm 2,010$	$8,752 \pm 1,920$	$24,860 \pm 1,030$
'Sepharose'-BSA	$1,514 \pm 260$	$1,236 \pm 110$	$1,474 \pm 461$
Uncoated 'Sepharose'	712 ± 190	814 ± 130	$1,061 \pm 260$

* All stimulants were titrated over a wide ($\sim 10^3$) range and only the maximal responses are given. (^{14}C)-Uridine and (^{3}H)-thymidine uptake data gave the same result in all experiments and therefore only the latter is presented. The numbers given represent the arithmetic mean plus or minus standard errors for replicate (three to six) cultures. The optimal amount of insoluble mitogen in terms of μg protein/ml. varied in the different experimental groups but in all cases was within 1 $\log_{10}$ of optimal concentration of soluble mitogen.
† Non-covalently bound PHA (see text).
‡ Optimal dose of 'Sepharose'-PHA (3) incubated for 48 h at $37°$ C in culture medium in the absence of cells. Beads spun down and supernatant medium added neat or 1 : 1 to freshly harvested lymphocytes.

PWM was purified from plant stems (*Phytolacca americana*) by the method of Borjeson *et al.*[13]. Purified phytohaemagglutinin (PHA), lot K 2402, was obtained from Wellcome Research Laboratories, Beckenham, UK, and bovine serum albumin (BSA) from Armour Pharmaceutical Co., Eastbourne, UK.

Packed agarose beads ('Sepharose 4B', Pharmacia) were diluted with an equivalent volume of distilled water, and activated by addition of saturated cyanogen bromide in a volume equal to that of the beads[14]. The beads were washed with 0.1 M $NaHCO_3$, pH 8.5, and then 1 mg of PHA, 10 mg of PWM or 20 mg of BSA was added per 1 ml. of packed 'Sepharose'. Mitogens were trace (1%) labelled with ^{125}I (1.6×10^5—5.4×10^5 c.p.m./mg). Ten times more PWM than PHA was added to give maximal stimulation of lymphocyte cultures. Unbound proteins were removed by washing with 0.1 M $NaHCO_3$ and phosphate buffered saline. The beads were incubated twice with sterile tissue culture medium for 24 h at $37°$ C. Less than 2% of ^{125}I-mitogen was released during subsequent incubation at $37°$ C. The degree of coupling of proteins to beads was determined by amino-acid analysis of their acid hydrolysates and by γ-radioactivity measurements: 60–75% of the PHA added, 67% of PWM and 20–25% of BSA were covalently conjugated to the 'Sepharose'.

To prepare 'B' cells 4-week-old CBA (H-2K) mice were thymectomized; 2 weeks later they were X-irradiated lethally with 850 r., and after a further 24 h reconstituted with 5–10×10^6 viable bone marrow cells pretreated with anti-θ serum (prepared and tested as described by Raff[28]) and guinea-pig complement to kill residual T cells. After 4–10 weeks, spleens from these mice contained less than 5% θ-positive cells and more than 74% MBLA-positive, cell surface immunoglobulin-positive (immunofluorescence), presumably B, lymphocytes[15,16]. T cells were obtained from thymuses of 4-week-old mice injected 48 h earlier intraperitoneally with 2.5 mg cortisone acetate (hydrocortisyl; Roussel, UK) to remove 90–98% of thymic cells leaving residual cells (all θ-positive) which were highly stimulated by phytomitogens[17] and possessed T cell immunological activity (for example, graft versus host capacity, and "helper" activity in antibody response[18]). Less than 1% gave positive immunofluorescent staining for cell surface immunoglobulin. Unselected B + T cells were prepared from spleens of 6 to 12-week-old normal CBA mice. Tissues were removed aseptically and teased into cell suspension in RPMI-1640 (Flow Laboratories) tissue culture medium supplemented with 10% heat inactivated foetal calf serum (Rehatuin, Armour Pharmaceutical Inc., Kaukakee), glutamine (4 mmol./ml.), penicillin (200 U/ml.), streptomycin (100 µg/ml.). All cell suspensions were filtered through cotton wool as previously described[19]. Eluted cells were over 98% viable lymphocytes and less than 1% demonstrably phagocytic (colloidal carbon uptake). Cells at 2×10^6/ml. concentration were cultured in the previously described culture medium for 24 or 48 h. Cultures were set up in microplates[19] (Cooke Ltd, Microtiter R) containing ninety-six wells with flat bottoms, each approximately 0.25 ml. capacity. They were cultivated at $37°$ C in a 7% CO_2, 10% O_2 and 83% N_2 atmosphere. To examine RNA synthesis, 0.05 µCi (^{14}C)-uridine (57 mCi/mmol., Radiochemical Centre, Amersham) was added to each culture after 20 h for a further 5 h. DNA synthesis was determined by adding 0.2 µCi (^{3}H)-thymidine (50 mCi/mol., Radiochemical Centre, Amersham) to each culture after 24 h for 16–20 h. The harvested cell suspensions were washed with cold phosphate buffered saline on to membranes (GF/C 2.5 cm, Whatman) in a Manifold multiple sample collector. Protein was precipitated by adding 5 ml. of cold 5% TCA, and the filters were washed with 5 ml. cold methanol. Membranes were dried, and then left in scintillation vials with 0.5 ml. solubilizer (Hyamine 10 x, Packard) at $45°$ C for 2 h. Then 5 ml. scintillation fluid (5 g PPO and 0.1 g POPOP in 1 l. toluene) was added to each vial. These were counted in a Beckman LS-200B liquid scintillation counter.

Table 2 Effect of Anti-θ (C3H) Serum on Phytomitogen-induced Proliferation

Serum *		Response: maximum uptake (counts × 10^{-3}/min) †	
		Unselected	
	Mitogen	(T + B) cells	B cells
Normal AKR	Control	$1,260 \pm 40$	$1,236 \pm 120$
serum	Soluble PHA	$21,045 \pm 1,990$	$1,409 \pm 260$
	'Sepharose'-PHA (3)	$36,142 \pm 2,640$	$34,862 \pm 2,100$
AKR anti-θ	Control	$2,140 \pm 2,040$	$1,410 \pm 110$
(C3H) serum	Soluble PHA	$2,253 \pm 1,860$	$1,636 \pm 120$
	'Sepharose'-PHA (3)	$32,468 \pm 2,105$	$36,480 \pm 1,850$

* 50×10^6 cells (in 2 ml.) treated with 1 : 4 serum at room temperature for 30 min. Cells washed and guinea-pig complement (1 : 8) added and cells incubated for a further 30 min at $37°$ C. Both cell suspensions were washed, cotton filtered and the viable cell count adjusted to 2.5×10^6/ml. for subsequent culturing.

material. This preparation bound 13% of the initial ^{125}I-labelled mitogen and induced a marginal, insignificant response *in vitro* ($1.3 \times$ control). Effectively all labelled mitogen could be eluted from these beads by N-acetyl-D-galactosamine. A similar treatment of covalently bonded PHA-'Sepharose' reduced the bound mitogen by only 5%. Perhaps the best evidence for the activating capacity of insolubilized mitogens derives from the experimental study itself in which B cells responded to 'Sepharose'-PHA, but were completely unresponsive to the soluble mitogen[19].

Implications for Lymphocyte Derepression

We conclude from these experiments than insoluble phytomitogens can activate both T and B lymphocytes and that in all probability this does not involve any internalization of the stimulant. This provides evidence that cellular derepression in this system is critically related to a plasma membrane triggering event and suggests that the same might be true for antigen. It is interesting to note that covalently linked antigens have been used to stimulate lymphocyte proliferation[20] and the release of migration inhibition factor (MIF) [20]. Although activation of cells by soluble ligands such as PHA may involve internalization of the stimulant[6], this can perhaps be viewed as a trivial consequence of pinocytosis induction at these areas of the membrane bearing the bound receptor sites. Alternatively, "ligand clearance" from the cell surface may be an important control mechanism modulating cell responses. In normal circumstances this would be achieved by pinocytosis. With insoluble stimulants detachment of cells from the ligand lattice, as observed in the current experiments, would be sufficient.

198

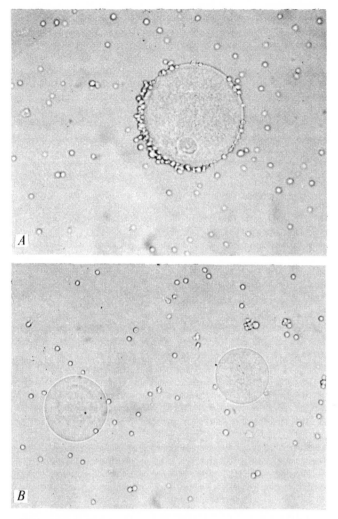

Fig. 1 Preliminary experiments to show that attachment to 'Sepharose' did not result in a loss of binding activity of the mitogen. Cells incubated with PHA-'Sepharose' for 60 min adhered to the surface of the 'Sepharose' particles (*A*) in contrast to cells incubated in the presence of BSA-'Sepharose' which remained free (*B*). ('Sepharose'-PHA was able to bind N-acetyl-D-galactosamine, as shown by its capacity to absorb the inhibitory activity of this sugar in binding and lymphoagglutination tests with soluble ^{125}I-PHA.)

The differential effect of soluble and insoluble PHA on B lymphocytes in this study is intriguing and may perhaps relate to antigen induced activation of these cells. There is increasing evidence that the stimulation of lymphocytes both by antigen and antisera to cell surface determinants involves cross-linkage or multi-point attachment of the stimulant[23,24]. Recent work (K. Lindahl-Kiessling, personal communication) with acid dissociated monovalent PHA suggests that this mitogen must also cross-link in order to stimulate cells. In addition, responsiveness of B cells to many antigens is dependent upon the activity of specifically reactive T cells[25].

One interpretation of this has been that T cells present antigen to B cells in a manner permitting simultaneous binding of multiple (identical) determinants[23]. These views gain credence from the observation that antigens with naturally repeating units (such as polymerized flagellin, pneumococcal polysaccharide, or heavily substituted hapten-protein conjugates) can stimulate B cells directly, or at least without a T cell contribution[24]. We can imagine that activation of B cells by 'Sepharose'-PHA is analogous to presenting these cells with a naturally polymeric antigen in that a greater and in this case obligatory degree of cross-linkage of stimulant bound to receptors is achieved than is possible with the soluble mitogen. This interpretation is supported by our preliminary observation that the density of phytomitogen molecules on the 'Sepharose' beads has a decisive influence on mitogenicity.

Similar reasoning has been used to explain the capacity of red cells to enhance the mitogenicity of PHA[26]. These results do, however, raise the question of why B cells, which have a similar average density of PHA binding sites as T cells[3,27], should presumably have a greater requirement for cross-linkage. This situation recalls the observation that B cells which probably have a higher density of antigen receptors than T cells[28,29] seem to be less readily stimulated or "tolerized" by antigen[30,31]. These paradoxical observations could perhaps be explained by specific requirements of B and T cells for triggering. Multi-point binding of stimulant will depend on a topographical correspondnece between the distribution of the receptors on the cell surface and the arrangement of reacting or determinant groups on the stimulant molecule, a variable independent of the average density of either. Binding sites for PHA on T and B cells might differ in this respect. In addition, if cell triggering involves, as is implied by our experiments, an initial plasma membrane phenomenon perhaps involving allosteric activation of enzyme systems, it is possible that T cells are either more richly endowed with the relevant plasma membrane associated enzymes. Another possible difference between T and B cells might be a differential sensitivity of their translocational/translational apparatus to activation or enhancement by cell surface stimulants; a difference which could be manifested at a variety of intermediate stages in the activation sequence.

We would like finally to draw attention to the possible

analogy between activation of lymphocytes by phytomitogens (and possibly antigens) and the behaviour of polypeptide hormones as cell stimulants. There is now compelling evidence that the receptor sites for these activating ligands are located on the plasma membrane[32,33] and that the inductive event concerns the activation of a membrane associated enzyme system such as adenyl cyclase[34]. It is also relevant to note that insoluble covalently linked insulin[35,36] and ACTH[37] both have biological activity *in vitro*.

The "second messenger" principle of hormone action expounded by Sutherland *et al.*[38] may well be operative in lymphoid cells, and indeed there is accumulating evidence that adrenergic receptor mechanisms and adenyl cyclase cyclic-AMP are involved in lymphocyte responses[39,40].

This work was supported by an EMBO fellowship (to S. B.) and the Medical Research Council.

[1] Allen, D., Auger, J., and Crumpton, M. J., *Exp. Cell Res.*, **66**, 362 (1971).
[2] Powell, A. E., and Leon, M. A., *Exp. Cell Res.*, **62**, 313 (1970).
[3] Greaves, M. F., Bauminger, S., and Janossy, G., *Clin. Exp. Immunol.* (in the press).
[4] Wigzell, H., *Transplant. Rev.*, **5**, 76 (1970).
[5] Raff, M. C., Sternberg, M., and Taylor, R., *Nature*, **225**, 553 (1970).
[6] Razavi, L., *Nature*, **210**, 444 (1966).
[7] Taylor, R. B., Duffus, W. P. H., Raff, M. C., and dePetris, S., *Nature New Biology*, **233**, 225 (1971).
[8] Han, S. S., and Johnson, A. G., *Science*, **153**, 176 (1966).
[9] Pogo, B. G. T., Allfrey, V. G., and Mirsky, A. E., *Proc. US Nat. Acad. Sci.*, **55**, 805 (1966).
[10] Killander, D., and Rigler, R., *Exp. Cell Res.*, **39**, 701 (1965).
[11] Frenster, J. H., *J. Cell. Biol.*, **43**, 39a (1969).
[12] Shyamala, G., and Gorski, J., *J. Biol. Chem.*, **244**, 1097 (1969).
[13] Borjeson, J., Reisfield, R., Chessin, L. N., Welsh, P., and Douglas, S. D., *J. Exp. Med.*, **124**, 859 (1966).
[14] Porath, J., Axen, R., and Ernbach, S., *Nature*, **215**, 1491 (1967).
[15] Raff, M. C., *Nature*, **224**, 378 (1970).
[16] Raff, M. C., Nase, S., and Mitchison, N. A., *Nature*, **230**, 50 (1971).
[17] Janossy, G., and Greaves, M. F., *Clin. Exp. Immunol.* (in the press).
[18] Blomgren, H., and Anderson, B., *Cell. Immunol.*, **1**, 545 (1970).
[19] Janossy, G., and Greaves, M. F., *Clin. Exp. Immunol.*, **9**, 483 (1971).
[20] Davie, J. M., and Paul, W. E., *Cell. Immunol.*, **1**, 404 (1970).
[21] Borberg, H., Yesner, I., Gesner, B., and Silber, R., *Blood*, **31**, 747 (1968).
[22] Amos, H. E., and Lachmann, P. J., *Immunology*, **18**, 269 (1970).

[23] Mitchison, N. A., *In Vitro* (Waverley, in the press).
[24] Feldman, M., and Basten, A., *J. Exp. Med.*, **134**, 103 (1971).
[25] Möller, G. (ed.), *Antigen Sensitive Cells*, in *Transpl. Rev.*, **1** (1968).
[26] Yachnin, S., Allen, L. W., Baron, J. M., and Svenson, R., *Proc. Fourth Ann. Leukocyte Culture Conference* (edit. by McIntyre, O. R.), 37 (Appleton Century Crofts, New York, 1971).
[27] Stobo, J. D., Rosenthal, A. S., and Paul, W. E., *J. Immunol.* (in the press).

[28] Raff, M. C., *Immunology*, **19**, 637 (1970).
[29] Jones, G., Torrigiani, G., and Roitt, I. M., *J. Immunol.*, **106**, 1425 (1971).
[30] Chiller, J. M., Habicht, G. S., and Weigle, W. O., *Science*, **171**, 813 (1971).
[31] Mitchison, N. A., in *Cell Interactions and Receptor Antibodies in Immune Responses* (edit. by Mäkelä, O., Cross, A., and Kosunen, T. U.), 249 (Academic Press, New York, 1971).

[32] Himms-Hagen, J., *Fed. Proc.*, **29**, 1388 (1970).
[33] Rodbell, M., Krams, M. J., Pohl, S. L., and Birnbaumer, L., *J. Biol. Chem.*, **246**, 1861 (1971).
[34] Pastan, I., and Perlman, R. L., *Nature New Biology*, **229**. 5 (1971).
[35] Cuatrecasas, P., *Proc. US Nat. Acad. Sci.*, **63**, 450 (1969).
[36] Oka, T., and Topper, Y. J., *Proc. US Nat. Acad. Sci.*, **68**, 2066 (1971).
[37] Schimmer, B. P., Veda, K., and Sato, G. H., *Biochem. Biophys. Res. Commun.*, **32**, 806 (1968).
[38] Sutherland, E. W., and Robison, G. A., *Pharmacol. Rev.*, **18**, 145 (1966).
[39] Hadden, J. W., Hadden, E. M., Middleton, E., and Good, R. A., *Intern. Arch. Allergy*, **40**, 526 (1971).
[40] Smith, J. W., Steiner, A. L., Newberry, W. M., and Parker, C. W., *J. Clin. Invest.*, **50**, 432 (1971).

B lymphocytes can be stimulated by concanavalin A in the presence of humoral factors released by T cells

J. Andersson , G. Möller and O. Sjöberg .

Abbreviations: Con A: Concanavalin A **LPS:** Lipopolysaccharide
T: Thymus-derived **B:** Bone marrow-derived

Concanavalin A (Con A) induces proliferation in thymus-derived (T) lymphocytes, whereas it does not induce DNA synthesis in bone marrow-derived (B) lymphocytes. However, in the presence of humoral factors released by normal thymus lymphocytes during 24 h in culture, B cells become competent to respond to Con A. The optimal dose of Con A for induction of proliferation in B cells treated with T cell supernatant was identical to that for T cell proliferation. Nonspecifically activated T cells were more effective than normal T cells to produce humoral factors making B cells competent to respond to Con A.

1. Introduction

Various mitogens vary in their ability to induce DNA synthesis in thymus-derived (T) and bone marrow-derived (B) lymphocytes. Concanavalin A (Con A) and phytohaemagglutinin (PHA) only stimulate proliferation in thymus lymphocytes and peripheral T lymphocytes [1–3] whereas other mitogens such as lipopolysaccharide from gram-negative bacteria (LPS) and poke weed mitogen have a selective stimulatory effect on B lymphocytes [1, 3, 4]. In spite of the high selectivity of the different mitogens with regard to activation of different cell types it has been shown conclusively that both T and B cells possess receptors for these different mitogens in approximately equal numbers [5].

Certain antigens require the cooperation of both T and B cells for the induction of antibody synthesis in the B cells [6, 7]. The mechanism by which T cells participate in the activation of B cell proliferation and differentiation into antibody-producing cells is not known, but three conceptually different hypotheses have been suggested. In one of these [8–10] it is postulated that the T cells bind the antigen molecules and present them to the B cells in such a form that the latter are activated, the idea being that B cells require a large number of antigen molecules being bound simultaneously to the antigen capturing receptors for activation to occur (local concentration). This is basically a quantitative proposal, implying that a sufficient energy of interaction must occur at the B cell membrane to achieve induction of proliferation in these cells. An alternative suggestion (for discussion see ref. [7]) is that the T cells activate the B cells by releasing humoral factors, which have the ability to make B cells respond to antigen. Finally, a molecular concept has been suggested [11], implying that the interaction between only antigen and receptor is a tolerance signal, whereas immunity requires the additional binding of an antibody to a different antigenic determinant on the antigen molecule.

Presently mitogens represent the best models for studying the mechanism whereby T and B cells become activated. We have previously demonstrated that the selective activation of T cells by Con A can be overcome if B cells are confronted with a local concentration of Con A molecules which was achieved by coupling Con A to the bottom of tissue culture petri dishes [12], thus lending support to the local concentration hypothesis. In the present experiments we tried to obtain experimental evidence for a competing hypothesis regarding B cell activation, namely that the activated T cells release factors, which act on B cells in such a way that they become competent to respond with proliferation after contact with the inducer, in our model being represented by Con A, which by itself is incompetent to activate B cells [3].

2. Experimental and Results

To achieve this we cultivated thymocytes for 24 h. There-after, the supernatants from these cultures were collected and added to spleen cells from congenitally athymic (nude) mice or to spleen cells from thymectomized, lethally irradiated mice repopulated with syngeneic bone marrow cells previously treated with anti-theta serum *in vitro* to eliminate functional T cells (T x B mice). These two types of spleen cell suspensions will be referred to as B cells. The B cell cultures were left untreated or were treated with various concentrations of Con A (Miles-Yeda Ltd., Rehovot, Israel, Lot no. 36) in the presence or absence of different concentrations of the supernatants mentioned above. The response of the B cells was determined after 72 h by incorporation of tritiated thymidine (1 μCi/ml, The Radio-chemical Centre, Amersham, England, spec. activity 5 Ci/mMol) added to the culture 24 h before harvest.

It was shown before that optimal stimulation of DNA synthesis in thymocytes by Con A is found after addition of 5 μg/ml, higher and lower doses giving a lower degree of thymidine incorporation. In contrast, Con A alone has no detectable effect on induction of DNA synthesis in B cells, as illustrated in Fig. 1 using spleen cells from nude mice. However, B cells can be induced to proliferate under these culture conditions by the addition of B cell mitogens, such as lipopolysaccharide from *E coli* bacteria (Fig. 1) purified from *E.coli* O55:B5 according to Westphal et al. [13, 14].

To investigate whether products from T cells can make B cells competent to respond to Con A, thymocytes from A x CBA mice were cultivated in tissue culture petri dishes (5 x 10^6 cells/ml) in Earle's solution containing 10 % fetal calf serum (Flow Laboratories, Irvine, Scotland). After 24 h, the cultures were harvested, centrifuged and the supernatant removed. This supernatant was added undiluted or diluted 1/2 with tissue culture medium to B cells derived from C57BL x DBA T x B mice. Some cultures received no further treatment, whereas others were given Con A in doses varying from 0.15 to 40 μg/ml. As can be seen in Fig. 2, Con A added to B cells in tissue culture medium only, did not stimulate DNA synthesis. Supernatants from thymus lymphocytes by themselves had no effect on the B cells in terms of cell proliferation. However, when the B cells were given both thymocyte supernatants and the various concentrations of Con A they responded with a marked increase in DNA synthesis. The dose response curve in this case was identical to that obtained with Con A stimulated T cells, 5 μg/ml being the optimal dose. When the thymocyte supernatant was diluted 1/2 the effect of DNA synthesis was the same, the dose response curve being identical to that obtained by undiluted supernatant. In control cultures, B cells were treated with the two different concentrations of thymocyte supernatant and at the same time with an optimally stimulating dose of lipopolysaccharide (LPS). As can be seen in Fig. 2, LPS caused a marked increase of DNA synthesis in these cultures as well as in cultures not receiving the supernatants. Thus, the supernatants by themselves do not stimulate DNA synthesis in B cells, nor do they change the B cell response to a B cell mitogen (LPS), but they make them competent to be activated by a T cell mitogen.

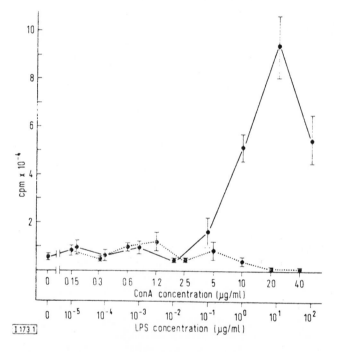

Figure 1. Response of nude spleen lymphocytes to various doses of Con A (●·····●) and LPS (●———●) as determined by incorporation of 1 μCi/ml of [³H]TdR on the 3rd day of culture.

Analogous experiments were performed with B cells obtained from the spleen of nude mice. As can be seen in Fig. 3, the 24 h supernatant derived from C57BL x CBA thymocytes had no stimulating activity itself on the B cells, whereas in the presence of Con A a typical dose response curve was obtained. Con A by itself had no effect on the B cells from nude mice.

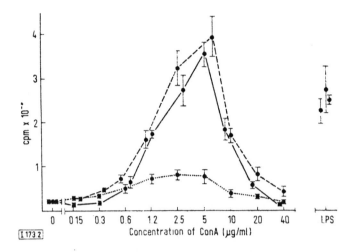

Figure 2. Effect of 24 h supernatant from normal A x CBA thymus lymphocytes on the response of C57BL x DBA T x B spleen cells to various doses of Con A. The response of T x B spleen cells to Con A in normal medium (● · · · · · ●), in undiluted (● — — — ●) supernatant and in supernatant diluted 1/2 (● ————— ●) is indicated. The response of T x B cells in the various media to $10\mu g/ml$ of LPS is indicated to the right.

A comparison was made between supernatants from untreated C57BL x CBA thymus cells and from thymus cells activated by Con A. The supernatants from such cultures were added to B cells (bone marrow cells from nude mice). As can be seen in Fig. 4, the supernatant from Con A stimulated thymocytes gave a stronger response than supernatants from untreated thymocytes. In this experiment the supernatant was taken after 48 h of cultivation, the medium being replaced after 24 h and cells treated for 60 min at 37 °C with 20 mg/ml of Methyl-α-D mannopyranoside (Calbiochem., San Diego, Calif.) in order to remove most of the Con A bound to the cells, whereas in the other experiments the supernatant was taken during the first 24 h.

Thus, supernatants from thymocytes made B cells competent to respond to Con A. The supernatant did not in itself induce DNA synthesis in the B cells. Con A by itself could not stimulate B cells at any concentration, but bound equally

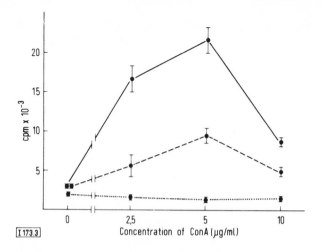

Figure 3. Effect of supernatants from 24 h thymus lymphocyte cultures upon the response of nude spleen cells to various doses of Con A. The thymocytes used for supernatant production were derived from C57BL x CBA mice pretreated with 125 mg/kg of cortison acetate 3 days before use, in order to remove cortison sensitive thymocytes [14]. Con A alone (● · · · · ·●), Con A and undiluted supernatant (●———●), Con A and supernatant diluted 1/2 in fresh tissue culture medium (● ——— ●) was added to the nude spleen cells.

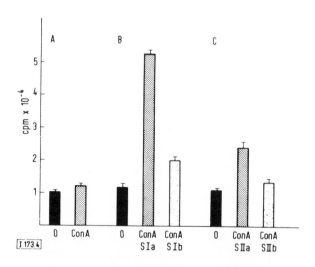

Figure 4. The effect of supernatants from normal (S II) or Con A stimulated (S I) C57BL x CBA thymus lymphocytes on the response of bone marrow (BM) cells from nude mice to 5 μg/ml of Con A.
A. Untreated or Con A treated BM in normal tissue culture medium.
B. Untreated BM in SI diluted 1/1.25 and Con A treated BM in SI diluted 1/1.25 (SIa) or 1/6 (SIb).
C. Same as in B but with supernatant SII.

208

to the B cells as to T cells. It is unlikely, therefore, that the supernatants from T cells act by increasing the efficiency of binding of Con A to the B cells. The mechanism by which the T cell factors and Con A operates to initiate DNA synthesis in B cells is not known, but it is obvious that the present results are compatible with the suggestion that the function of T cells in the normal immune response to T cell dependent antigens is to release a factor making the B cells competent to respond to antigen by proliferation and antibody synthesis, which the cells cannot do in the absence of this factor. This finding together with the previous demonstration that locally concentrated Con A causes B cells to initiate proliferation, although the degree of the response was smaller to locally concentrated Con A than to Con A plus the T cell supernatant, suggests that there are at least two pathways by which B cells can be activated. If one T cell function is to release factors making the B cells competent to respond to inducing agents (mitogens or antigens) it follows that T and B cell cooperation in the induction of antibody synthesis is to a certain extent non-specific. This is an apparent contrast to *in vivo* findings demonstrating specificity of the T and B cell interaction. However, if the T cell factors act over a short distance only and if both T and B cells must interact with the antigen, specificity would be preserved.

The part of this work carried out in Sweden was supported by grants from the Swedish Medical Research Council, the Swedish Cancer Society, the Wallenberg Foundation and the Harald Jeansson's Foundation. The technical assistance of Miss Monica Gidlund and Miss Birgitta Hagström is gratefully acknowledged.

Received November 27, 1971.

3. References

1 Janossy, G. and Greaves, M.F., *Clin. Exp. Immunol.* 1971. *9*: 483.

2 Greaves, M.F. and Roitt, I.M., *Clin. Exp. Immunol.* 1968. *3*: 393.

3 Andersson, J., Möller, G. and Sjöberg, O., *Cell Immunol.,* in press.

4 Peavy, D.L., Adler, W.H. and Smith, R.T., *J. Immunol.* 1970. *105*: 1453.

5 Greaves, M.F., Bauminger, S. and Janossy, G., *Clin. Exp. Immunol.,* in press.

6 Transpl. Rev. 1969. *1*:

7 Miller, J.F.A.P. and Osboa, D., *Physiol. Rev.* 1967. *47*: 437.

8 Mitchison, N.A., *Cold Spring Harbor Symp. Quant. Biol.* 1967. *32*: 431.

9 Möller, G., *Cell. Immunol.* 1970. *1*: 537.

10 Taylor, R. and Iversen, G.M., *Proc. Roy. Soc. Lond.* B. 1971. *176*: 393.

11 Bretscher, P. and Cohn, M. *Science* 1970. *196*: 1042.

12 Andersson, J., Edelman, G.M., Möller, G. and Sjöberg, O., Eur. J. Immunol. 1972. *2*: in press.

13 Westphal, O. and Jann, K., in "*Methods in Carbohydrate Chemistry*", Academic Press, New York, 1965. *5*: 33.

14 Westphal, O., Lüderitz, O. and Bister, F., *Z. Naturforsch.* 1952. *7b*: 148.

15 Andersson, B. and Blomgren, H., *Cell. Immunol.* 1970. *1*: 362.

KEY-WORD TITLE INDEX

AUTHOR INDEX

Agarossi, G., 12, 124
Allison, A.C., 68
Andersson, J., 203

Basten, A., 161
Bauminger, Sara, 194
Benezra, D., 27
Breitner, J.C.S., 53
Burg, C., 172

Cheers, Christina, 53
Cudkowicz, Gustavo, 72

Davies, A.J.S., 68
Dennert, G., 109
di Pietro, S., 12, 124
Doria, G., 12, 124

East, June, 140

Feldman, Michael, 17, 98
Feldmann, Marc, 60
Folch, Hugo, 10

Gery, I., 27
Globerson, Amiela, 17
Greaves, Melvyn F., 178, 194

Janossy, G., 178

Kettman, John R., 81
Kleinman, R., 27
Kunin, Sara, 17

Lennox, E., 109
Little, Margery, 53
Lonai, Peter, 98

Mäkelä, O., 151
Miller, J.F.A.P., 53, 161
Mitchison, N.A., 32
Möller, Erna, 151
Möller, G., 203
Morgado, Fernando, 10

Niederhuber, J., 151

Orsini, Frank R., 72

Playfair, J.H.L., 89

Raff, Martin C., 128, 140
Roelants, G.E., 167

Sabolovic, D., 173
Schimpl, A., 118
Segal, Shraga, 17
Shearer, Gene M., 17
Sjoberg, O., 203
Sprent, J., 161
Sulitzeanu, D., 27

Vann, Douglas C., 81

Waksman, B.H., 140
Wecker, E., 118
Wilson, J.D., 60
Wioland, M., 173

Guide to Current Research

The research summaries appearing in the following section were obtained through a search of the Smithsonian Science Information Exchange data base conducted in July, 1973.

The Exchange annually registers 85,000 to 100,000 notices of current research projects covering a wide range of disciplines and sources of support. SSIE endeavors to retain up to two full years of current research information in its active file. The selection of summaries appearing in this section does not represent the complete SSIE collection of information on this topic, but, rather, has been specifically tailored to reflect the scientific content of this particular volume. A limited number of summaries may have been omitted because clearance for publication by the supporting agency or organization was not received prior to the publication date.

SSIE is the only, single source for information on ongoing and recently terminated research in all areas of the life, physical, behavioral, social and engineering sciences. The SSIE file is updated daily by a professional staff of scientists utilizing a comprehensive and flexible system of hierarchical indexing. Retrieval of subject information is conducted by these same specialists, using computer-connected, video display terminals which allow instant access to the entire data base and on-line refinement of search strategies. SSIE offers an information service unequalled anywhere: comprehensive and vital information on who is conducting what research where and under whose support.

More current information, and in some cases expanded coverage, on the topic considered in this volume is available directly from SSIE. This information is offered at modest cost in the form of custom searches of the SSIE file designed specifically to meet the user's need or as an update of the subject search in this section. For more information on SSIE, contact MSS or write directly to the Smithsonian Science Information Exchange, 1730 M Street, N.W., Washington, D.C. 20036. Subject search or updated package requirements may be discussed with SSIE scientists by calling the Exchange at (202) 381-5511.

CONTROL MECHANISMS IN LYMPHOID
DIFFERENTIATION,
R. AUERBACH, Univ. of Wisconsin, Graduate
School, Madison, Wisconsin

The characterization of some of the
interactions involved in the development of
immunity in vertebrates as well as in the
regeneration of immune competence after
x-irradiation is being undertaken. The
demonstration of intrinsic differentiation
of thymic lymphoid cells, the behavior of
thymic grafts, and the ontogenetic pattern
of lymphopoiesis (formation of the white
blood cells) led to the suggestion that the
thymus may be the primordium of the immune
system.
    The major question facing the group is
what is the source of the immune competent
cells? Experiments are underway to test
critically the role of the yolk sac, liver
and thymus cells in immunological
maturation. The assay systems to be used
will inclue both the ability to evoke
graft-versus-host response and the ability
to produce specific antibodies. While the
thymic role in origin of immune competent
cells will be examined, analysis of the
thymic activity observed to occur across
millipore filters will also proceed. The
systems currently available for study are
the regeneration of immune competence in
irradiated spleen, and the stimulation of
lymphopoiesis in bone marrow cultures.
Efforts will be made to determine the
ontogeny of and the cell type associated
with the production of specific factors.
Finally, studies are contemplated using
several mutant mouse strains with
interesting histological effects on the
thymus.

IMMUNOCOMPENTENCE OF THYMUS CELLS,
H.N. CLAMAN, Univ. of Colorado, School of
Medicine, Denver, Colorado 80220

This project investigates the immune
response, particularly the kinds of
lymphocytes involved in making antibody. We
have established that both thymus-derived
and bone-marrow-derived cells are involved.
We will investigate the activities of these
cells in the thymus, the marrow and the
peripheral lymphoid tissue. We will look at
the turnover of cells, their response to
antigens and nonspecific cell mitogens.
There appears to be a differential effect of
corticosteroids on thymus-derived and
marrow-derived cells, depending on their
state of maturation and location. This will
be investigated using measurements of
antibody production and cell-mediated
immunity. Species differences in the
response of lymphoid cells will be studied.
The induction of tolerance to haptens on
various carriers will be looked at.

BASIC STUDIES OF THE IMMUNE RESPONSE AND THE
EFFECT OF RADIATION ON THIS RESPONSE,
F.J. DIXON, Scripps Clinic Res. Fdn., San
Diego, California 92037

Studies of human glomerulonephritis
will concentrate on the nature of the
glomerular basement membrane antigens which
are responsible for the anti-GBM antibody
formation underlying one form of this
disease. The search for possible etiologic
agents of immune complex type
glomerulonephritis will focus on E-B virus
and hepatitis virus, both of which have been
shown to have some relationshp to human
glomerulonephritis.

The effect of radiation and chemically induced immunosuppression on the course of non-cytocidal chronic viral infections will be observed. These studies should help differentiate between any direct deleterious effects of the presence of virus and any injury caused by an immune response to the virus.

The effect of radiation on T and on B lymphocytes will be observed both before and after the exposure of these cells to antigen. These studies should reveal any difference in radiosensitivity of these two types of lymphocytes and their precursors and demonstrate any increased radioresistance gained by these cells after antigenic stimulation.

The events contributing to the induction of unresponsiveness in spleen cells exposed in vitro to antigen for four hours will be analyzed. In this situation the roles of both thymus and bone marrow cells can be established and any interaction of these two cell types recognized.

Results: Chemical analysis of human GBM membrane has been carried out. Evidence has been obtained that the EB virus may be important in a significant number of patients with glomerulonephritis. A technique was designed to measure the number and type of immunoglobulin molecules on the surface of living lymphocytes. Further studies were carried out on the immunopathology of chronic viral infections and the mechanisms involved in autoimmune-like phenomena in New Zealand mice.

STUDIES IN THE CONTROL OF PROLIFERATION AND DIFFERENTIATION IN THE IMMUNE RESPONSE, R.W. DUTTON, Univ. of California, Graduate School, San Diego, California 92038

We are engaged in a systematic study of

the cellular events involved in the immune
response and the ontogeny of immunocompetent
cells. We are using an in vitro culture
system of mouse spleen cells. We are able to
measure the contributions of thymus-derived
cells (T cells) and bone marrow-derived
cells (B cells) to the response by using
trinitrophenyl-conjugates of heterologous
erythrocytes as antigen. We are
investigating the regulation of the immune
response and the interactions of cellular
and humoral responses. We will investigate
ways to manipulate these interactions to
favor tumor rejection.

IMMUNE RESPONSES TO TUMOR GRAFTS,
R.K. GERSHON, Yale University, School of
Medicine, New Haven, Connecticut 06510

    The role of thymus-derived (T.D.)
lymphocytes in the immune response to tumor
grafts will be studied. The DNA synthetic
response of T.D. cells to tumor grafts, in
terms of magnitude and in terms of
proportion of dividing cells will be
studied, in both lymph nodes and spleen.
The concomitant immune response, in terms of
cell mediated immunity and antibody
production will be correlated with the
response of T.D. lymphocytes.
    Dose kinetic studies will be done. The
effect of splenectomy and tumor excision on
the response will be ascertained. The
results with tumors will be compared with
those produced by normal cells.
    The functional role of T.D. lymphocytes
will be investigated in allogeneic chimeras
by transferring their spleen cells to
appropriate iso-immune, irradiated secondary
hosts and observing the effect of these
cells on tumor growth. Normal cells will be
added to the secondary hosts also, to study

the need for cell to cell co-operation.

The effects of T.D. cells on target cells in tissue culture will be studied.

The specificity of the T.D. cell response will be investigated by labelling those that respond to a first tumor injection with H3 thymidine. Comparison of % labelled cells responding to a second tumor injection of the same or different tumors will then be made.

FUNCTIONAL ONTOGENY OF THE IMMUNE RESPONSE, E.S. GOLUB, Purdue University, School of Science, Lafayette, Indiana 47907

The purpose of this research is to investigate the functional ontogeny of the immune response. In the past several years it has become evident that several cells are involved in the immune response and the interrelationships between these cells in the immunologically mature animal are becoming clear. The precursor of the Antigen Sensitive Cell has been identified in the mouse as a bone marrow derived cell which has reacted in some manner with a thymus derived cell. Antigen is probably processed by a macrophage and presented to the Antigen Sensitive Cell in such a manner that this cell proliferates and differentiates into an Antibody Forming Cell. Thus, at least three cells have been identified and are thought to interact. The purpose of the proposed research is to investigate the development of the ability of the cells concerned to participate in the generation of Antigen Sensitive Cells.

In vitro stimulation of mouse, Rana and Xenopus spleens will be carried out using tissues from animals of progressively younger ages with a variety of antigens. It is postulated that interacting cells in the system do not develop simultaneously and by determining the earliest age at which the system is complete Dr. Golub will have a starting point in investigating the development of the various components. Use of in vitro methods permits manipulation that use of the whole animal prohibits.

As a second approach, the age at which the thymus gains the ability to participate in the generation of Antigen Sensitives Cells will be determined. This will be done by using embryonic thymus to study the marrow-thymus interaction and the ability of thymus to regenerate competence to irradiated spleen populations.

PATTERN OF HEMOPOIETIC DIFFERENTIATION IN MARROW-PLUS -LYMPHOCYTE CHIMERAS, J.W. GOODMAN, Oak Ridge National Laboratory, Oak Ridge, Tennessee 37830

Many interesting data have been obtained from microscopic examination of splenic colonies in marrow and thymocyte chimeras. One approach was to study the relationship of thymus-marrow synergism to GVH was to substitute lymph node lymphocytes for thymocytes and then to look for augmentation. Using Fe59 uptake as an endpoint, at least 8 such experiments were negative. However, when we look microscopically at the spleens of chimeras that had received lymphocytes but were not taking up Fe59, we found some granulocytic colonies. It seemed desirable to explore this histologic finding further.

Results: A preliminary experiment confirmed (by Fe59 uptake and gross surface spleen colonies) the lack of augmented erythropoiesis in lymphocyte-marrow chimeras, but microscopic counts of colonies showed that augmentation of marrow growth had taken place, and most of the additional colonies were granulocytic. We are currently investigating this apparent change in differentiation pattern.

LYMPHOCYTE INTERACTIONS, RECEPTORS AND FUNCTIONS,
I. GREEN, U.S. Dept. of Hlth. Ed. & Wel., P. H.S. Natl. Insts. of Health, Bethesda, Maryland 20014

A variety of different studies are being performed: A. A receptor on bone marrow derived (B) lymphocytes for antigen antibody complement complexes has recently been described. We have found that the lymphocytes from 8/8 patients with chronic lymphatic leukemia have this receptor. Also, cells from a number of human lymphoblastic cell lines also possess these receptors. Therefore, these leukemic cells and some of the human lymphoblastic cell lines are derived from B cells. B. There are several types of lymphocyte mediated cytotoxicity. In one type normal lymphocytes in the presence of antibody cause destruction of the target cell. Spleen cells from x-irradiated thymectomized, bone marrow reconstituted mice are just as effective as cells from intact mice. Prior exposure of spleen cells to anti-theta serum and complement do not change the cytotoxic ability of these cells, whereas exposure of spleen cells to anti-kappa serum, with or without complement, markedly impairs the ability of these cells to cause cytotoxicity. These data suggest that a B lymphocyte or subclass of B lymphocyte is responsible for this antibody dependent cytotoxicity. C.

Lymphocytes proliferate in vitro in response to antigen. This is felt to be an in vitro manifestation of cellular immunity and to be mediated by thymus derived (T cells). However, some data suggest that B cells may also proliferate in these cultures. To determine the precise degree of T cell proliferation of lymph node lymphocytes in vitro in response to antigen, complement receptor lymphocytrosettes (which form only on B cells) were done in cultures to which 3H thymidine has been added. Then by radioautography, the cells that formed rosettes that also had incorporated 3H thymidine could be determined. About 30% of proliferating cells in these cultures formed rosettes and are therefore B cells.

CELLULAR INTERACTIONS IN THE ANTIBODY RESPONSE,
D.L. GROVES, Wake Forest University, School of Medicine, Winston Salem, North Carolina 27103

The overall goal of this program is to contribute to an understanding of the mechanism of interaction between thymus-derived helper cells (T cells) and bone marrow-derived precursor cells (B cells) in the antibody response of mice to sheep erythrocytes. During the past year the equivalence of limiting dilution analysis and the hemolytic focus assay has been demonstrated indicating that these techniques measure the same entity. Studies have been conducted in order to characterize a priming phenomenon occurring in recipient mice, most of which are not able to make a

221

primary antibody response due to the limited number of spleen cells transferred. Maximum priming for direct PFC occurs in 5-6 days and the process appears to involve cell multiplication, possibly of T cells. The degree of priming is inversely related to the dose of antigen employed in the fist injection. The frequency of the units which become primed is about twice that for direct PFC response units in the normal spleen. Data for indirect PFC are erratic and difficult to interpret. Previously unanticipated studies have commenced on the mechanism of antigen competition and are being carried out in an experimental model involving diffusion chambers. Work continues on the secondary antibody response in diffusion chambers.

IN VITRO ISOGENEIC LYMPHOCYTE INTERACTION, M.L. HOWE, State University of New York, School of Medicine, Brooklyn, New York 11203

The purpose of this project is to pursue an extensive study of a newly described in vitro phenomenon; the Isogeneic Lymphocyte Interaction (ILI). It is proposed that the ILI which is characterized by proliferative activity in cultures containing neonatal mouse thymocytes mixed with isogeneic adult spleen cells represents an in vitro model, the study of which will aid in determining the function of the thymus in the development of cellular immunity and the mechanisms by which lymphoid cell recognition and discrimination occur within the body. Experiments have been designed, using in vitro tissue culture methods as well as manipulation of lymphoid cells in living animals to identify the antigens which trigger the ILI; the cause of the sudden appearance of ILI activity in the

222

neonatal thymus; the distribution of ILI
activity in other species; and the
relationship of the ILI to other in vivo
phenomena.

LYMPHOCYTE KINETICS AND CHARACTERIZATION,
L.I. JOHNSON, Univ. of The West Indies,
Kingston, Jamaica

     This research will study proliferation
as a characteristic of cellular immune
responses and a function of the immunologic
competence of the lymphocyte. The cell
culture approach will be employed, since its
proliferation characteristics resemble the
cellular response occurring in
graft-versus-host (GVH) and homograft
reactions in vivo.
     The specific aims of the research will
be: a) to analyze the proliferation
kinetics of lymphocytes (immunologically
normal and abnormal), using the cell cycle
kinetics as a possible means of
distinguishing subpopulations of
lymphocytes. Further characterization of
cells will be based on specific functional
tests such as evoking GVH reaction and
acting as multipotential colony-forming
units. Suicidal H3T will be employed in
combination with the double isotope
technique for determination of cell cycle
parameters of specified cell groups. Bone
marrow lymphocytes will be separated on
density gradients and grouped according to
differences in their proliferation kinetics,
and functional performance. These studies
relate directly to the problem of bone
marrow transplantation and GVH disease.
They may aid in the development of a

technique for suppressing undesired cellular
responses in transplantation and autoimmune
diseases. They also hold promise of
benefits in the characterization and
diagnosis of certain forms of leukemia
having disorders of lymphocyte
proliferation.

GENETICS AND REGULATION OF ANTIBODY
FORMATION,
E.S. LENNOX, Salk Inst. For Biolog. Studies,
San Diego, California 92109

Work has continued on the functions of
thymus derived lymphocytes. In particular
comparison has been made of their role in
cooperating with bone marrow derived
lymphocytes to induce the humoral response
as well as their role in cell mediated
immunity. It has been possible by using
several different cellular antigens as
carrier for haptens or as targets for cell
mediated attack to dissociate these two
activities. We have paid special attention
to cell mediated immune reactions in which
antibody plays a role. We are continuing to
analyze systems in which the thymus derived
lymphocyte may be itself the cell that
carries out lytic attack on the target.
    The goal of this is to understand
better the interplay of components in the
immune response to tumors.

STUDIES OF CELL SURFACE ANTIBODY ON
THYMUS-DERIVED AND BONE MARROW-DERIVED
IMMUNOCOMPETENT CELLS,
J. LESLEY, Salk Inst. For Biolog. Studies,
San Diego, California 92109

    In the work being proposed here, I will

be studying immunoglobulin- like molecules on the surface of thymus-derived (T cells) and bone marrow-derived (B cells). These molecules are believed to be antigen receptor molecules involved in specific recognition of antigen as a first step in the initiation of an immune response or of immunological tolerance. The points to be studied are: 1) Isolation of the T cell receptor, characterization and function. 2) dynamics of T and B cell receptors at the cell surface. 3) Significance of receptor turnover in the processes of driving the cell to immunity or tolerance. 4) Relation of cell surface immunoglobulin to other lymphocyte surface antigens.

AGING AND IMMUNE COMPETENCE - IN VITRO CHARACTERIZATION OF IMMUNE DEFICIENCY IN AGED MICE,
T. MAKINODAN, Oak Ridge National Laboratory, Oak Ridge, Tennessee 37830

In vitro studies have shown that the observed humoral immune response impairment in old mice is associated with the nonadherent cell population which contains thymus-derived lymphocytes (T-cells) and bone marrow-derived lymphocytes (B-cells). To investigate possible causes of the deficiency in this cell population, spleen cell suspensions from young and old mice are separated into distinct populations, initially by their ability to adhere to plastic and subsequently by their density in Ficoll gradients. The identity of the cell types present in the various fractions is determined with anti-sera and mitogens specific for B- or T-cells. The cell fractions are cultured in various combinations with sheep red blood cells, and their antibody-forming capacities are assessed.

    Results: In the Ficoll gradient, the
nonadherent cells separate into two
fractions, a light fraction rich in B-cells
and a dense fraction rich in T-cells. The
separation profile of cells from young and
aged mice is distinctly different.
Recombining the old cell fractions in the
ratio observed in the young did not improve
the response obtained from the old spleen
cells. Hence, the deficiency in the old
mice is not due to a shift in the ratio of
the essential cell types but apparently to
the impaired function of B-cells and/or
T-cells.

STRUCTURE AND FUNCTION OF CELL SURFACE
IMMUNOGLOBULIN,
J.J. MARCHALONIS, Walter & Eliza Hall
Institute, Melbourne, Victoria, Australia

    This project is directed towards
elucidating the role of cell surface
immunoglobin in immune reactions carried out
by lymphocytes. Two major classes of
lymphocytes occur; namely, thymus-derived
lymphocytes (T-cells) and
bone-marrow-derived lymphocytes (B-cells).
Both cell types possess specific surface
receptors for antigen; however, only the
B-cell line is active in secretion of
antibodies. T-cells mediate cellular immune
reactions including graft-rejection and
suppression of tumor growth. Furthermore,
these cells cooperate with B- cells in the
generation of an antibody response to many
antigens. This research proposal will
investigate the structural properties of
isolated surface immunoglobulin, the
receptor for antigen, from T-cells, B-cells
and tissue culture lines of T-cell tumors.
The isolation of lymphocyte surface

226

immunoglobulin will be carried out using lactoperoxidase- catalyzed radioiodination of living cells. In addition to immunoglobulin, cell surface structures such as the theta antigen and the cell coat will be analyzed. The binding specificity of isolated surface immunoglobulin from T-cells activated specifically to proteins, erythrocytes, histocompatibility antigens and tumor antigens will be ascertained. Moreover, in vitro experiments designed to ascertain the mechanism of collaboration between T cells and B cells will be carried out. These studies will analyze the function of specific complexes of cell surface immunoglobulin with its antigen and also possible roles of other surface components.

LYMPHOCYTE IN IMMUNITY - EMPHASIS ON SURFACE IMMUNOGLOBULIN FUNCTION,
R.E. MASS, U.S. Veterans Administration, Hospital, Portland, Oregon 97207

A study of the presence of surface immunoglobulins on lymphocytes, the probable recognition site of antigen on the "memory" cell is to be carried out in patients with immunocyte malignancy, autoimmune disorders and on fetal thymic cells.
Fluorescent conjugated anti-immunoglobulin will be used to identify IgM, IgA, IgG and light chain on the lymphocyte surface.
Significance: Evidences of an increase in a specific "B" lymphocyte identified by immunoglobulin specificity on surface lymphocyte is of interest in relating the type of lymphocyte involved in immunocyte malignancies and in the autoimmune disorders. Studies on fetal thymic cells are of interest in evaluating the presence

of surface immunoglobulins in "T" cells
since the fetal thymus persumably contains
no "B" cells.

Progress: Initial progress was slowed
because of difficulty in obtaining
equipment, and in the completion of the
laboratory facilities, however, in the
summer of 1972 studies were begun to
evaluate the reliability of surface
immunoglobulin identification by the
fluorescent antibody technique. To date a
series of patients with chronic lymphocytic
leukemia had been done as controls and the
technique is satisfactory using Meloy
laboratory commercially available
fluorescent tagged anti-immunoglobulin.
Studies on fetal thymus cells have been held
up because of the lack of fetal thymus. We
are currently beginning evaluation of
lymphocytes in autoimmune disorders.

THYMUS GRAFTS AND TUMOUR INDUCTION,
J.F. MILLER, Walter & Eliza Hall Institute,
Melbourne, Victoria, Australia

The thymus channels the differentiation
of hemopoietic stem cells to thymus
lymphocytes of which some emigrate to
populate the thymus- dependent areas of the
lympoid tissues and to recirculate from
blood to lymph. These thymus-derived
lymphocytes or T cells are involved in
cellular immunity - including killer
activity against many types of tumour cells.
Nonthymus-derived lymphocytes or B cells
are antibody- forming cell precursors and,
in the case of tumours, can produce
cytotoxic antibodies, but, more often,
enhancing or blocking antibodies which
protect the tumor against the killer
activity of T lymphocytes. Ways have been
found to selectively activate T lymphocytes

against specific cellular antigens and to
selectively deplete lymphocyte populations
from B cells. The aim of this research
project is to devise more effective methods
of activation of T cells against tumor
specific antigens and to repress the ability
of B cells to produce blocking antibodies.
Numerous in vitro and in vivo systems are
being tested in mice with a view to
achieving such an aim.

ROLE OF ANTIBODIES AND CELLULAR INTERACTIONS
IN DELAYED HYPERSENSITIVITY AND LYMPHOCYTE
TRANSFORMATION,
J.J. OPPENHEIM, U.S. Dept. of Hlth. Ed. &
Wel., P.H.S. Natl. Insts. of Health,
Bethesda, Maryland 20014

Mechanisms by which different cell
populations involved in immunological
reactions interact are being investigated.
The role of thymic dependent T cells and B
cells that produce antibodies, in in vitro
lymphoproliferative reactions and in vivo
manifestations of delayed hypersensitivity
is being studied utilizing
agammaglobulinemic chickens (depleted of B
cells) and chickens treated with ALS
(depleted of T cells). The facility with
which macrophage-bound antigens induce
delayed hypersensitivity and activate thymic
dependent T rather than B types of
lymphocytes is also being studied using
similar approaches to obtain selected
populations of lymphocytes. Antibodies made
by B cells have been found to inhibit the
lymphoproliferative reactions of T cells in
vitro. The role of this apparent feedback
control in alleviating symptoms of patients
with allergies that are undergoing treatment
with hyposensitization is being studied.

229

CELL INTERACTIONS IN IMMUNE RESPONSES IN
VITRO,

C.W. PIERCE, Harvard University, School of
Medicine, Boston, Massachusetts 02115

The purpose of this project is to
investigate the nature of the essential
interactions among cells in the immune
system that are required for development of
immune responses in vitro. Interactions
among antigen, macrophages, thymus-and bone
marrow-derived cells in cell clusters are
necessary for development of IgM, gamma 1,
gamma 2, and IgA antibody responses. The
morphological relationships among these
cells in clusters and the function of each
cell type is under intense investigation.
An in vitro system for maturation and
differentiation of thymocytes from
immunologically unreactive to reactive,
competent cells has been developed and will
be used to analyze maturation of parameters
of thymus cell function. The effects of
non-specific activation of thymus-derived
cells by phytomitogens on their capacity to
mediate reactions of cellular and humoral
immunity will be analyzed. Lastly, factors
influencing the shift in immunoglobulin
receptor from IgM to IgG class is under
intense investigation.

IN VITRO STUDIES OF ANTIGEN RECOGNITION AND
PROCESSING,
S.F. SCHLOSSMAN, Harvard University, Beth
Israel Hospital, Boston, Massachusetts 02215

The isolation and characterization of
defined populations of immunologically
competent lymphocytes and their products
should allow investigations designed to
elucidate the mechanism of antigen

recognition. We plan to prepare cellular immunoabsorbants using dextranase digestible insoluble support of Sephadex which will allow both depletion and recovery of populations of functionally viable lymphocytes according to their surface properties. Specific T cells, B cells and null cells will be investigated either separately or in recombined pools both in vivo and in vitro to elucidate their role in the induction of the immune response. The technique for fractionation of lymphoid cells into functionally viable populations with specific receptors for antigen will be utilized in attempts to define the chemical nature, the mode of attachment, distribution and specificity of the receptors on both T and B cells. Moreover, we plan to undertake genetic studies of the immune response which take advantage of the limited heterogeneity of antibodies and sensitive measures of T and B cell responses to defined DNP-oligolysines. With the use of isoelectric focusing to define antibody responses in clonal terms and the development of antibodies to these clones we intend to: 1) define the cellular site for immune gene response control; 2) ascertain the V gene pool size to defined antigens; 3) segregate those genes governing antibody specificity from those regulating the magnitude of the immune response and 4) investigate to defined DNP-oligolysines.

THE CELLULAR SEQUENCE OF IMMUNITY, M.R. SCHWARZ, Univ. of Washington, School of Medicine, Seattle, Washington 98105

The general research objective of the investigator is to delineate the cellular sequences of immunity as they relate to

immediate and delayed hypersensitivity
states, passive transfer of immunity,
homograft reactions (including graft versus
host reactions) and tolerance. Since three
cell lines (the lymphocytic, plasmocytic and
phagocytic) have been implicated as playing
major roles in such immune responses,
investigative efforts have been concerned
with elucidating the origin, life span,
fate, function and interrelationships of
these cell types. Of especial interest at
the present time is the development of in
vitro methods for the detection of temporary
and permanent states of tolerance as they
relate to homograft immunity, relation of
thymus to tolerance and the relationship
between the synthesis of RNA and
immunological competence.

CELL MODEL FOR THE INITIATION OF ANTIBODY
RESPONSES,
S. STROBER, Stanford University, School of
Medicine, Palo Alto, California 94305

The objective of the proposed
investigation is to test the following
hypothesis: a) Long-lived, non-dividing,
recirculating lymphocytes can initiate the
secondary antibody response in rats. Some
of these cells are "non-thymus derived" and
are the precursors of the antibody forming
cell. Others are "thymus derived" and
augment the response produced by the
"non-thymus derived" cells by binding
antigen and presenting it efficiently to
them. b) Short-lived, non- recirculating,
"non-thymus derived" lymphocytes are the
precursors of the antibody forming cells in
the primary response. "Thymus derived"
cells similar to those above also augment
this response. c) In order to study this
model, purified populations of the types of

232

lymphocytes described above will be tested for their ability to restore the adoptive primary and secondary antibody responses of X-irradiated rats to proteins and to hapten-protein conjugates.

CHEMICAL AND SEROLOGICAL STUDIES OF LYMPHOCYTE SURFACES,
J.L. STROMINGER, Harvard University, School of Arts, Cambridge, Massachusetts 02138

The objective of this work is to distinguish subpopulations of lymphocytes on the basis of chemical differences of their cell surfaces. The domestic fowl will be used as a source of lymphocytes for this purpose because of the well-defined delineation of the immunological system in this species. Chickens contain a bursa-dependent system which is mainly responsible for the humoral immunological defense mechanism, (for example, that against bacterial infections) and a thymus-dependent (cell-mediated) system which is mainly involved in graft rejection, delayed type hypersensitivity, and the graft versus host reaction. By bursectomy or thymectomy respectively, in combination with whole body irradiation, one or the other system can be depressed.
    Different subpopulations of lymphocytes, distinguished on the basis of chemical and serological differences would thus be correlated with one or the other functional system. After these subpopulations have been clearly distinguished, we wish to isolate the surface substances responsible for these differences and to characterize them by structural analysis.

CELL INTERACTIONS IN ANTIBODY FORMATION, D.W. TALMAGE, Univ. of Colorado, School of Medicine, Denver, Colorado 80220

We are attempting to separate, characterize, assay and determine the function of the three cells required for the antibody response to sheep erythrocytes. The current working hypothesis is that the auxiliary cells (A cells) represent a third universe of clones, each making a distinct mediator. In regard to antibody formation the two most important types of A cell would be that which makes an inducer of B cells and that which kills B cells. The function of the antigen specific thymocyte is to activate its A cells. The function of adjuvants such as poly AU and endotoxin is to enhance the cell interaction. Various aspects of this model are being tested by direct experiment.

SYNERGY IN THE GRAFT-VERSUS-HOST RESPONSE, R.E. TIGELAAR, U.S. Dept. of Hlth. Ed. & Wel., P.H.S. Natl. Insts. of Health, Bethesda, Maryland 20014

Graft-versus-host (GVH) reactions are produced by inoculation of mouse lymphoid cells from various tissues either alone or in various combinations into newborn F1 recipients. Reactions are quantitated by the Simonsen spleen weight assay, in which the degree of spleen enlargement is a linear function of the logarithm of the cell inoculum. Previous studies have indicated that optimal expression of GVH reactions may involve interaction among at least two subpopulations of thymus- derived cells. Further studies are required to characterize these subpopulations and elucidate the nature of this interaction. Either normal or thymectomized donors of lymphoid cells or such cell suspensions in vitro will be treated with a variety of agents with affinity for cell membrane constituents (such as anti-thymocyte serum, anti-theta, anti-TL, cortisone acetate) and the cells then examined for alteration in their behavior in GVH reactions. Two extensions of the basic project involve: (1) testing the capacity of various cell populations alone and in combination to cause rejection of SV-40-induced tumors in mice, and (2) determining if the cell responsible for amplification of GVH reactions has physiologic properties similar to the thymus-derived "helper" cell involved in many humoral antibody responses.

LYMPHOCYTE FUNCTIONAL CAPACITIES, W.T. WEBER, Univ. of Pennsylvania, School of Veterinary Medicine, Philadelphia, Pennsylvania 19104

Experiments are designed to characterize the functional differences of lymphocytes of the bursa of Fabricius and thymus in chicks and to determine the potential of embryonic marrow stem cells to develop into a functional thymus-derived or bursa-derived lymphocyte population. The investigations employ various techniques e.g. the transfer of sex chromosomally marked bone marrow or bursal cells to appropriate normal, agammaglobulinemic or thymectomized animals followed by in vivo and in vitro procedures designed to test the functional capacity of the transferred cells. Special emphasis is on the response of thymus- derived and bursal and bursa-derived cells to various specific

antigens, phytomitogens, and to allogenic
cells.

ROLE OF THE THYMUS IN IMMUNOLOGICAL
DEVELOPMENT,
I.L. WEISSMAN, Stanford University, School
of Medicine, Palo Alto, California 94305

This project is involved with the
generation of thymus derived immunologically
competent and active cells.   Three lines of
approach are used: 1)   Testing for tumor
analogues of thymus derived cells, in order
to obtain large populations of homogeneous
(cloned) cells for detailed study of T cell
surface receptors for antigens.   2)
Development and purification of antisera
specific for lymphocyte subpopulations of
the B cell (bursal) and T cell (thymic)
lineage; 3)   Use of these antisera to aid in
definition of the various morphological
substructures in peripheral lymphoid
tissues, as well as aiding the definition of
T cell and B cell functions in cellular
immunity.